FANS

FANS

Avril Hart and Emma Taylor

V&A Publications

First published by V&A Publications, 1998
V&A Publications
160 Brompton Road
London SW3 1HW

Avril Hart and Emma Taylor assert their moral right to be identified as
the authors of this book

Designed by Polly Dawes

ISBN   1851772138

A catalogue record for this book is available from the British Library

Printed in Hong Kong

JACKET ILLUSTRATION
*Front and back: French fan, designed for the Spanish market,
1820s-1830s. Engraving, painted in gouache on paper,
showing front and back views of fashionably dressed ladies
and a gentleman. Sticks of painted wood, resembling guitars.*
T.123-1920

FRONTISPIECE
Detail of plate 47: French or Dutch silk brisé fan, 1820-30s.

CONTENTS

INTRODUCTION

The telegraph of Cupid in this fan
Though you should find, suspect no wrong;
Tis but a simple and diverting plan
For Ladies to chit-chat and hold the tongue.

(From 'The Original Fanology or Ladies' Conversation Fan', a fan leaf designed by Charles Francis Bandini and printed by William Cock, 42 Pall Mall, and Robert Clarke, 26 Strand, in 1797.)

The fan has been an essential female dress accessory for several centuries, evolving from simple beginnings to an exquisitely decorated artefact over a period of at least four hundred years. It has also been instrumental in furthering discreet social intercourse between the sexes, acquiring the power to communicate sentiments from flirtatious desire to devoted love through its ever more sophisticated use of imagery. The language of the fan developed properly with the folded or pleated fan, which reached the height of perfection in the eighteenth century, and no doubt served as an amusing diversion, providing delicious opportunities for mischievous flirtation.

The fascination that fans exert for collectors and non-collectors alike is at least twofold: as attractive ornaments in themselves, and as examples of skilled craftsmanship whose development is linked with the history of the decorative and fine arts. Collectors have as many reasons to acquire fans as there are types of fan. Some will concentrate on a particular period or style; others might collect a single type of fan – the brisé or printed fan, perhaps, or fans from China or Japan. This selective approach is of course purely personal, but is an indication of the variety of different types of fan and the interests they arouse and satisfy. A famous nineteenth-century fan collector, Lady Charlotte Schreiber, collected mainly printed fans and fan leaves, the majority of which reflected her interests in history, politics and social events. The important collection she built up was presented to the British Museum.

Once you begin to look at fans, questions arise about where, how and when they were made, about how current affairs or fashion affected their design, about what kind of materials were used and why, and about who made them and who bought them. For instance certain fans of the seventeenth and early eighteenth centuries stimulate an interest in classical mythology and the history of painting, while others send us back to the history of wars and revolutions, by which their imagery was inspired. It was in the eighteenth century that fan manufacture and design not only reached new levels of perfection, but also became a thriving industry. Fashion decreed that every woman should have a fan, and these ranged from cheap printed souvenirs bought on the street for a few pence when going to the theatre, to exquisite examples worth hundreds of pounds with a superbly painted leaf and intricately carved ivory sticks inset with jewels. The refinement of printing techniques and paper manufacture at that time also paved the way for the political fan, which could be produced at speed and sold on the streets before the news it announced became stale. With the tremendous changes in fashion that occurred in the nineteenth century, fan manufacture began to experience a decline, although the fans of this era are some of the most exciting and varied in design: all the elegance, vigour and vulgarity of the century is reflected in their decoration.

This book reveals some of the finest examples of the Victoria and Albert Museum's superb fan collection, exploring the fascinating history of European folding fans from their heyday in the seventeenth and eighteenth centuries through to exceptional examples from our own. Those described and illustrated in the following pages range in

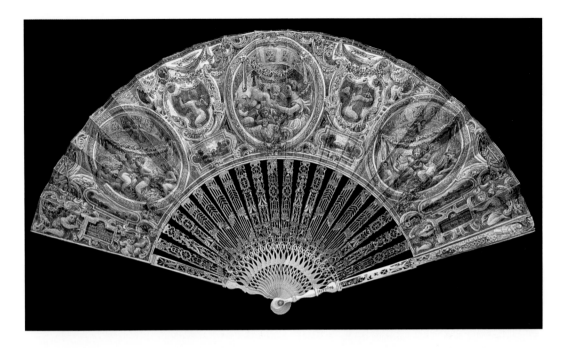

Plate 1 'Allegories of Love'. Gouache on vellum, by Sir Matthew Digby Wyatt. English, 1869. 2177-1876.

style and technique from a rare 1620s brisé fan of appliquéd straw work from Italy to an equally unusual fan of painted silk, designed and painted by Duncan Grant for the Omega Workshops in London in 1913.

The V&A's collection of European fans has primarily been formed as a result of the acquisition by gift or bequest of several outstanding private collections. The first and most significant gift of over 400 fans was made as early as 1876 by Sir Matthew Digby Wyatt (1820-77) and Lady Wyatt, whose high-quality collection also included fans from China and Japan. (Until 1975, when a separate Far Eastern Department was formed at the Museum, European and Far Eastern fans were both collected by the Textiles Department.) Sir Matthew Digby Wyatt was an eminent artist and art historian, as well as acting as an Art Referee for the Museum, one of a small group of experts 'selected from the most competent authorities' and appointed by a committee to advise on purchases. Plate 1 shows a fan illustrating

'Allegories of Love', designed and painted by Sir Matthew as a gift to his wife, and is signed and dated 1869. Sir Matthew's magnificent bequest forms the nucleus of the Museum's collection and has remained the largest independent gift of fans to the V&A. It comprises a wide range of styles and types from the late seventeenth to the nineteenth century, and is particularly rich in both printed and painted examples from the eighteenth century.

Several other major gifts or bequests have been made over the subsequent hundred years, contributing to a collection that illustrates a wide variety of popular subjects in various techniques. Classical, romantic, pastoral, religious and domestic scenes are all depicted on fans that display a breathtaking range of craftsmanship and artistry, as they evolved to keep pace with the perennial taste for novelty. Prior to these bequests, the Museum had already begun to acquire a small number of superior-quality fans from the major exhibitions held in the second half of the nineteenth century, such as the Great Exhibition of 1851 and the various International Exhibitions in Paris. The fan featured in plate 55, for instance, was purchased from the Paris International Exhibition of 1867, and was painted by Edouard Moreau (1825-78) for the French firm Alexandre.

This book sets out to illustrate the development of fan design and the variety of techniques used in their production. It is not intended to explore their development chronologically, but to show the range of fan design over a period of four centuries through a variety of different themes, focusing either on choice of subject or technique. However, to set the themes in context – which include chinoiserie, brisé, tourist and commemorative fans – we start with a brief history of the European folding fan.

CHAPTER I

BEGINNINGS: THE SIXTEENTH AND SEVENTEENTH CENTURIES

Folding fans with an attached pleated leaf originated in Japan from about the twelfth century AD, but did not reach Europe until the second half of the sixteenth century. They were most probably received at first as intriguing curiosities, but their design would not have appealed to fashionable European taste at the time, which favoured elaborate decoration on every surface. Japanese design was as rigorously restrained then as it is today, with no superfluous ornament; their fans then were not so different from those of the next four hundred years. The sticks were of narrow bamboo, some of which were lacquered in black and may also have had gilt decoration, and the leaves would have been of paper, bearing minimal decoration, perhaps a Japanese inscription. Nevertheless, the principles of Japanese fan construction were not lost on European craftsmen. The shape (which opened out to a quadrant of a circle), the number of sticks, and the placing of the rivet which held the sticks together all seem to have inspired them to produce fans along similar lines. Although very few European folding fans from this period have survived, they can be seen in many portraits of the period, especially from the 1580s onwards. The majority are much more elaborate than contemporary Japanese fans, with distinctive European decoration. These luxurious accessories would have been exclusive to rich and aristocratic society, while ordinary people would have continued to use cheap examples such as the flag fan shown in fig.2.

The book *Habiti Antichi di tutto il Mondo*, published in 1589 by Cesare Vecellio, is illustrated throughout with wood-cuts showing contemporary fashionable dress worn by citi-

MATRONA FERRARESE.

Fig.1 Matrona Ferrarese holding a folding fan, from *Habiti Antichi di tutto il Mondo*, by Cesare Vecellio, 1589. NAL 2058-1880.

zens of different countries. The majority of the women hold folding fans, most of which appear to be of European design (fig.1). Until their introduction, Europeans had used various kinds of handscreen, which were rigid fans attached to a handle. One type consisted of a short handle inset with feathers; another was a flag-shaped fan with a long handle and a flap like a flag attached to one end (fig.2). A few flag fans survive and these are made of dyed and intricately woven straw, palm or reed leaves or of découpé vellum.

One of the earliest surviving European folding fans, from the late sixteenth century, is held in the collection of the Musée de la Renaissance, Ecouen, France. It has roughly carved bone sticks, its outer or guard sticks are decorated with little tufted pom-poms of cerise silk, and the leaf is of leather (vellum). The leaf is attached to the sticks using a somewhat crude method of sliding them through a series of small radiating horizontal slits; the effect is rather like sliding a hand into a glove, except that the sticks are allowed to show in between the slits. The leaf also bears rows of small painted mica insertions like little windows, which turned the fan into a fascinating ornamental toy. The introduction of mica as a decorative material was a relative novelty in sixteenth-century Europe. Tufted silk pom-poms, however, were a popular decoration for the outer sticks – they can be seen on the sticks of the closed folding fan held by Elizabeth I (1558-1603) in the Ditchley portrait of c.1592 by Marcus Gheeraerts the Younger (1561-1636) in the National Portrait Gallery, London. A similar fan is referred to in the inventories of Elizabeth I for 1590: 'new Tuftinge of a Fanne of perfumed Lether', and again in 1596, when several of the queen's folding fans were repaired: 'tufting parte of them with carnation ingrayne silke, for glewinge & revetting them, part with revettes of gold, and parte of silver, with one ounce quarter di [1 3/8oz] of pearle [probably picots of silver-gilt thread around the edge of the leaf] to perfourme the edginge'. Elizabeth I's inventories were published in their entirety by Janet Arnold in *Queen Elizabeth's Wardrobe Unlock'd*.

The brisé is an alternative type of folding fan which con-

Fig.2 Meretrici Publiche holding a flag fan, from *Habiti Antichi di tutto il Mondo*, by Cesare Vecellio, 1589.
NAL 2058-1880.

11

sists of a set of decorated sticks without a pleated leaf. (The term 'brisé' has only been applied to this type of fan since the early twentieth century.) One of the earliest representations of a brisé fan is in a portrait of Princess Elizabeth (1596-1662), daughter of James I, painted when she was a child in 1603 and now in the National Maritime Museum at Greenwich. She is shown holding a closed fan in her left hand. In this example, the fan appears to be made from about six leaf-shaped panels of découpé vellum, each leaf inserted onto a straight, narrow stick. The outer stick is decorated with seven tufts or pom-poms of white silk.

The earliest fan in the V&A's collection is a brisé fan (plate 2). Probably Italian, it dates from the 1620s and has seven sticks, each one shaped to represent a curled ostrich feather. The 'feathers' are made of rigid grey paste board, reinforced with fine metal rods or wires and covered in green

Plate 2 Brisé fan. Paste board covered in silk, decorated with straw appliqué work. Italian, 1620.
T.184-1982.

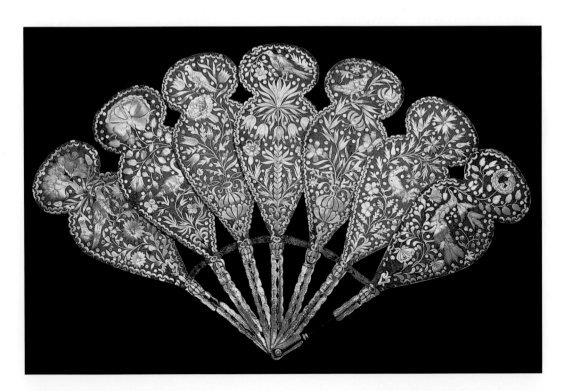

silk decorated on both sides with high-quality appliquéd straw work; this work may have been done in Florence, one of the centres for this craft in the early seventeenth century. The sticks are linked together by a green silk ribbon which allows the leaves to overlap when the fan is open. Feather handscreens were one of the most popular types of fan in the sixteenth century and these brisé ones may have been the only way to achieve a folding feather fan. Several portraits painted in the 1660s by Girolamo Forabosco (1604-79) show fans of a similar shape to the one in the V&A, and a further example can be found in the Bayerisches National Museum, Munich. Folding fans using real feathers do not appear until the second half of the nineteenth century, although feather handscreens continued to be popular.

Construction techniques employed by European makers of folding fans in the late sixteenth and early seventeenth century appear to be rather clumsy. It seems they were experimenting with different materials, including embroidered silk, leather and paste board, which often proved too heavy for the sticks. Some ivory sticks appear to have been quite substantial with high relief carving, and may well have become too heavy for the leaf. The craftsmen had to hone their skills to suit the scale of the object, for fans at this time were not large, being approximately eight to nine inches (about 23 cms) long, and opening out to about sixteen to eighteen inches (40 to 50cms). Their comparatively small size meant that elaborate leaves or sticks would have made them awkward to handle.

During the seventeenth century, however, techniques improved. The sticks, which could be of ivory, bone or tortoiseshell, became much thinner and those that supported the leaf were exceptionally fine and tapered to a point. The guard sticks became broader and blade-like so that the pleats of the fan leaf were protected when the fan was closed. The guards had minimal decoration in the form of low relief carving or incised motifs. Some were inlaid with mother-of-pearl and tortoiseshell or, later, decorated with silver piqué. The technique of silver piqué had been intro-

Plate 3, overleaf 'Venus and Cupid'. Gouache on leather.
French, *c.*1670-80.
T.184-1920.

duced in Naples in the last quarter of the seventeenth century and remained fashionable until the first quarter of the eighteenth century. Examples of the highest quality came from Italy or France, also centres of finely carved high-quality ivory fan sticks. The improvements in technique enabled the leaf to vary in size, allowing the entire structure to expand from a quadrant shape to a semi-circle. Some of the largest semi-circular fans spanned as much as two feet.

Greater use was made of pleated fan leaves of thin leather, painted in gouache. Many late seventeenth- and early eighteenth-century leaves were of a dark purplish-brown leather (vellum). A dark ground allowed the painted colours to glow with a greater intensity. Fan painters were inevitably influenced by the masters of the Early and High Baroque, copying or adapting their paintings for fan leaves. Many artists who reached maturity of style and technique at this time were pupils of the Carracci family, who founded a teaching academy in Bologna in c.1585, or were followers of Michelangelo Merisi da Caravaggio (1573-1610). His use of chiarascuro in particular influenced painters throughout the seventeenth century, and his strong contrasts of light and dark, brilliance and deep rich shadows are often faithfully reproduced on fans of this period.

The literary and artistic tastes of the era too were reflected in the choice of subject: classical, biblical and to a lesser extent conversation scenes predominated. The quality of the painting was often very high, as in the exquisite fan leaf of 1670-80 depicting Venus with putti (plate 3). The reclining Venus and putti occupy a central cartouche surrounded by superbly painted flowers which glow against the dark ground colour of the fan leaf. The flowers include those associated with Spring and themes of love and betrothal.

The reverse side of the fan leaves was usually painted with sprays or bunches of flowers, often of great beauty and again painted with great skill and attention to botanical detail. Another common feature of the reverse was a serpentine line roughly traced down the sticks in gold, silver, or some other colour on an unpainted ground.

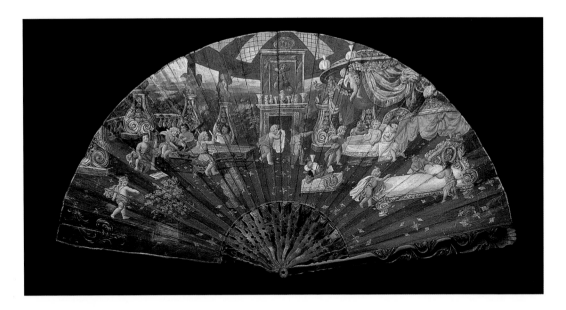

A popular subject for the front of the leaf was 'The Toilette', which along with the theme of 'The Five Senses' illustrated the contemporary association between love and luxury. The fan of the 1670s-80s depicted in plate 4 is one of the larger fans, opening to a full semi-circle and mounted on tortoiseshell sticks. A large and opulent chamber opens on one side to a view of a distant landscape. Within the richly furnished room, putti are preparing Cupid's bath. This setting is very similar to that of a series of paintings on the theme of 'The Five Senses' by Jan Brueghel the Elder (1568-1625), painted in 1618, at a time of growing awareness of and sensual pleasure in luxury goods. Each painting shows a vast room open on at least one side and flanked with great columns or arches through which a distant landscape can be seen. Objects associated with each of the senses are carelessly scattered around the room in rich profusion.

A very grand and high-quality fan of the 1670s, thought to represent Madame de Montespan (1641-1707) at her toilette (plate 5), clearly displays this taste for excessive consumerism. The famous mistress of Louis XIV sits in an imaginary cham-

Plate 4 'The Toilette of Cupid'. Gouache on leather, tortoiseshell sticks.
French, 1670s.
T.155-1978.

ber which combines aspects of the Apartement des Bains and the Trianon de Porcelaine at Versailles, where she is surrounded by her attendants amidst luxurious furnishings and exotic objects, while putti prepare her bath and cool the air with long feather fans. The semi-circular leaf was converted into a rectangular painting, perhaps for a gentleman's cabinet of curiosities. The style and technique of the fan painter has been so faithfully retained in the additional areas that the shape of the original fan leaf is almost indistinguishable.

George Woolliscroft Rhead in his book *History of the Fan*, published in 1910, comments that 'the fan representing the "Toilette of Madame la Marquise de Montespan" and the "Promenade" in the possession of the Countess Duchatel has become historic'. Rhead claims that the fan was originally sent by Madame de Sévigné (1626-96) to her daughter, Madame de Grignan (1646-1705), and that it was referred to in a letter of 1671: 'My fan has then become most useful, doubtless. Do you not think it beautiful?' Unfortunately

Plate 5 'Toilette of Madame de Montespan'. Gouache on leather. French, 1670s. P.39-1987.

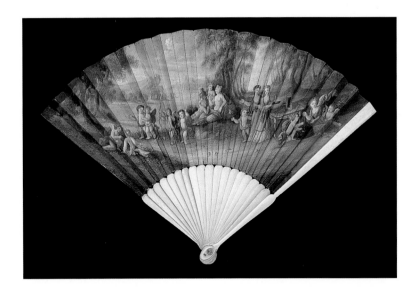

Plate 6 'Toilette of Venus'. Gouache on leather, plain ivory sticks.
French, 1690-1710.
T.162-1920.

Rhead does not describe or illustrate this fan in his book. It is tempting to speculate if the converted fan leaf in plate 5 is the same as that in the possession of Madame de Sévigné. It would certainly have provoked comment at the time; the opulent furniture is similar to items designed for Louis XIV, and the mirror and overmantel are each surmounted by the king's sun symbol. A fan of this quality must indeed have belonged to someone at the court of the Sun King.

Plate 6 is a complete contrast to the opulence of the previous fans, setting the 'Toilette of Venus' in a landscape beside a lake with figures dressed in the fashions of the period. The leaf is of dark brown leather painted in gouache, and the sticks, though plain, are of the finest ivory. This is a typical example of a good-quality fan of the late seventeenth and early eighteenth centuries.

The majority of fans to survive from the seventeenth century date from the last quarter, and are usually of high quality with well painted leaves and richly decorated sticks. They were obviously costly, and would have been preserved with care as valuable heirlooms to be handed down from one generation to the next.

CHAPTER II

CLASSICAL MYTHS

Some of the most outstanding fan paintings have been inspired by the myths of the classical world. Their more immediate source, however, was the contemporary master-pieces of the artists of the Early and High Baroque.

Two of the finest and most interesting fans in the V&A collection both illustrate the same subject of 'Venus and Adonis' and have been taken from the same painting by Francesco Albani (1578-1660) in the Villa Borghese, Rome. The mount-ed fan with tortoiseshell sticks of 1680-1720 seen in plate 7 is painted in watercolour on kid using a fine stipple tech-nique and is signed 'Leonardo Germo of Rome'. Its minia-turist painting technique is characteristic of the finest Italian fan paintings of the time: the artist has considerable control using this method and can create a highly modelled and detailed finish. This particular fan once belonged to the American painter Benjamin West (1738-1820), who arrived in Italy in 1760 to study for three years before making his home in London. He was to become President of the Royal Academy in 1792.

Germo's fan can be compared with the fan leaf of the 1720s illustrated in fig.3, which is drawn in pen and ink on kid leather and signed 'Domenico Spinetti. Napolitano. Roma. F'. The drawing has been carried out in minute detail and with very little resort to washes, most of it being done by hatching and stippling. The composition is almost identi-cal to the fan by Germo and is obviously taken from the same painting, although the grouping has been reversed, and there is an additional seated figure in the foreground. Fans are very rarely signed and the fact that both of these have an identifiable artist's signature is of great assistance in

Plate 7 'Venus and Adonis', by Leonardo Germo of Rome. Watercolour on kid leather. Tortoiseshell sticks.
Italian, early 18th century.
2200-1876.

Fig 3 'Venus and Adonis', by Domenico Spinetti. Pen and ink on kid leather, Italian *c*.1720.
E.1024-1970

assigning a convincing date, or at least a period and provenance for each. Normally dating has to be judged by the style and subject matter of the painting and composition. The sticks are not necessarily a useful guide to this, as their fragility required them to be frequently replaced. The duplication of subjects on fans is of interest to the historian, as it shows how popular a particular subject may have been over a period of time.

Another Italian fan, of *c.*1700-20, is shown in plate 8. It is after the painting by Guido Reni (1575-1642) from the Casino Rospigliosi, Rome, and is from the same Early Baroque period as the painting of Venus and Adonis. The original work is faithfully and delicately reproduced on the small scale of a fan leaf. Painted in watercolour on kid, it is another example of the Italian genius for fine fan painting using the

Plate 8 'Aurora and the Hours with the Chariot of the Sun', after Guido Reni .
Watercolour on kid leather, ivory sticks. Italian, early 18th century.
65-1870.

pointillist technique. The ivory sticks are carved and pierced and inset with tortoiseshell around the rivet. The Cooper-Hewitt Collection, New York, has an almost identical fan leaf. There is an alternative version of Aurora painted by Guercino (1591-1666) which shows her in a chariot driving away Night, which appears on a fan of the same period in New York's Metropolitan Museum. The survival of these fans in such widespread and public collections, even though they can be counted on the fingers of one hand, is evidence of the popularity of the theme they illustrate at the time they were produced.

The High Baroque is superbly represented by 'The Triumph of Alexander', seen in the fan of 1670-1700 illustrated in plate 9. This fan shows specifically his entry into Babylon, and is taken from one of a set of five paintings on the History of Alexander executed between 1661 and 1668 by Charles Le Brun (1619-90) for Louis XIV. Le Brun was appointed as director of the Gobelins factory in 1663 and

Plate 9 'The Triumph of Alexander'. Gouache on leather, ivory sticks decorated in silver piqué.
French, 1670-1700.
2276-1876.

23

also supplied designs for tapestries, one of which was a set of 'The Triumph of Alexander'. As a young man he had worked for a time with Nicolas Poussin (1594-1665) and absorbed his theories on classicism in art, principles which he continued to endorse when he became director of the Académie in 1663. They also found expression in his painting, which formed the basis not only of French academicism but of the Grand Manner of Louis XIV.

The fan painting is an almost identical copy of the original by Le Brun in the Louvre. The artist has cleverly adapted the composition to fit the curved shape of the fan leaf by shifting the groups of figures slightly and introducing a completely new group in the centre foreground, in front of Alexander's chariot. He has also changed the balance of colour by introducing strong cobalt blues. In the original painting the blues are much lighter and Alexander is wearing a gold-coloured cloak, whereas on the fan it is a rich blue. Alexander's helmet is also of gold and it bears the laurel wreath of the victor. On the fan the helmet has been given

Plate 10 'Venus interceding with Jupiter'. Gouache on paper, carved ivory sticks decorated with mother-of-pearl.
French or English, mid-18th century.
T.85-1956.

a red plume. It was not uncommon for artists to have to use engravings of great works of art as their source of design, as they were often unable to gain access to the originals. The changes in colour on this particular fan could be due to the fact that the artist may have only had a black and white engraving of the painting as a reference. The reverse is painted with flowers. The sticks, which are contemporary with the leaf, are of ivory and richly decorated with silver piqué and inset with mother-of-pearl around the rivet. There is another version of this painting on an early eighteenth-century fan in a private collection, where red has become the predominant colour for the clothes and the composition has again been slightly altered by the introduction of additional figures.

Eighteenth-century fans inspired by classical mythology tended to attract a very high standard of painting. The fan showing Venus interceding with Jupiter (plate 10), and another with Bacchus and Ariadne as its subject (plate 11),

Plate 11 'Bacchus and Ariadne'. Gouache on leather, carved ivory sticks.
English or Dutch, mid-18th century.
531-1869.

both date from the middle of the eighteenth century and could be of French or English workmanship. They are of the highest quality, where both the sticks and the fan leaves are of an equal standard, and they would have been costly items. Each artist has his own individual style, suggesting that neither came from a particular studio where the style had become stereotyped.

The fan showing Bacchus and Ariadne (plate 11) is painted in subdued colours, but has the additional decoration of appliquéd straw work amongst the flowers at either side of the leaf, giving an added dimension when the straw glitters like gold as the fan is turned in the light. This would have been particularly effective in candlelight. The sticks are of ivory, richly carved and pierced. Plate 10 has a powerfully painted leaf showing Venus interceding with Jupiter, possibly on behalf of Aeneas, her son. The ivory sticks are not

Plate 12 'Hector's Farewell'. Gouache on paper, carved mother-of-pearl sticks.
English or Dutch, 1730-40s.
T.87-1956.

only carved and pierced but also inlaid with chips of mother-of-pearl: this too would have sparkled.

A finely painted fan showing 'Hector's Farewell' (plate 12), from Homer's *Iliad* (Book XXII), copies the earlier painting of the same subject by Antoine Coypel (1661-1722). This beautiful artefact has, once again, superbly carved and pierced mother-of-pearl sticks, which again would have added to the richness and luminosity of the entire fan.

The fans discussed in this chapter are of exceptionally high quality, and as such represent the jewels in the crown of the art of fan painting. For it is these early, fine fans (which occur in all good fan collections) that set a standard of painting and craftsmanship that was rarely surpassed in later periods. The choice of classical subjects taken from great works of art helped to set the standard of design, while the painting was carried out by professional artists. These fans are individual items, intended for or possibly commissioned by a rich and cultured clientele. However, the increasing popularity of the fan lead to greater numbers being produced in the latter half of the eighteenth century, and this gradually brought about a decline in quality which continued into the nineteenth century.

CHAPTER III

ELYSIAN FIELDS: EIGHTEENTH-CENTURY PASTORAL FANS IN ENGLAND AND FRANCE

The scenes of quiet flirtation and summer walks in the country which are characteristic of eighteenth-century pastoral fans might be seen to echo the activities of the ladies carrying them. Fans were associated with dalliance throughout the eighteenth century, and writers often refer to the 'language' or secret messages presumed to be conveyed by ladies with their fans. It is unlikely that any very complex code could ever be agreed and conveyed to a lover, but women were certainly aware that their fans could be used as a weapon in attracting men. In May 1764, Cleone Knox recorded preparations for a party in her diary:

> My sister, examining me, was highly satisfied with my appearance, only admonishing me to use my fan gracefully, for said she, 'There is a whole Language in the fan. With it the woman of fashion can express Disdain, Love, indifference, encouragement and so on'. To tell the truth I had never thought of all this before, having found my eyes sufficient up to now to convey any message I wished to the other Sex.

In 1711, the *Spectator* satirized the use of fans in flirtatious performances by ladies. It was claimed that 'women are armed with Fans as Men with Swords, and sometimes do more Execution with them'. The author then imagines an academy in which ladies might be trained to 'Handle your Fans', 'Unfurl your Fans', 'Flutter your Fans' and so on. He particularly comments on the pleasant moment when the fan is unfurled and suddenly 'an infinite number of Cupids, Altars, Birds, Beasts, Rainbows and the like agreeable

Figures...display themselves to View, whilst every one in the Regiment holds a Picture in her Hand'.

The pastoral images found so often on fans were really visions of escape to the countryside for wealthy people, adapted by fan painters from the art and literature popular at the time. The term 'elysium' or 'elysian fields' derives from classical mythology and was often applied to these imagined landscapes. While we often know very little about the individuals who painted these scenes, the artistic influences upon their work are easy to trace.

The French artist who most influenced the painters of pastoral fans throughout Europe was Antoine Watteau (1684-1721). Some of his most famous paintings – *La Conversation* (1712-13), *Les Plaisirs du Bal* (1716-17) – in which small groups of men and women talk earnestly and intimately together, provided fan painters with poses and compositions ideal for reproduction in miniature. All these works have meadow or garden settings, and Watteau's masterpiece, *Pélégrinage à l'Isle de Cythère*, shows men and women in fashionable clothes escaping in boats to a remote Greek island dedicated to Venus, Goddess of Love. Such scenes appealed both to French courtiers detained at Versailles and unable to visit their estates, and to the English aristocrats' love of the countryside.

What is perhaps surprising is that few scenes painted on fans show a woman actually using her fan, whereas in Watteau's *Les Plaisirs du Bal*, for example, several women play coyly with fans as they listen to the flattering words of ardent young lovers. Some gaze so intently at their fan that they appear quite oblivious to the young men kneeling at their feet. Fans had many uses in company. They could cover embarrassment, as in Watteau's painting; alternatively, many eighteenth-century fans had 'peep holes', through which a lady could observe the object of her flirtation. Fans could cover a lack of conversation, according to M. de Saussure who visited England in the 1720s and found that 'the main conversation is the flutter of fans'. They also had certain ceremonial functions at the French court, where it

Plate 13 Gouache on vellum.
French, *c.*1760.
2190-1876.

was customary for ladies to present gifts to the queen, held on an open fan.

The subject and style of the fan in plate 13, which dates from around 1760, is clearly influenced by Watteau. A group of prosperous young men and women have escaped to an isolated meadow to eat, drink, and play music. The empty bottles and hastily folded picnic rug in the foreground suggest that the meal is over, the musician has just started playing, and the most inebriated member of the party is dancing, bottle and glass in hand. Sheep carry on munching the grass. The air of innocence and the surprising lack of flirtatious behaviour in the scene give this fan its charm, although one of the women in the centre seems to be admiring the stockinged legs of the musician!

The other French artists whose styles were emulated by fan painters are Francois Boucher (1703-70) and Jean-Honoré Fragonard (1732-1806). Watteau, Boucher and

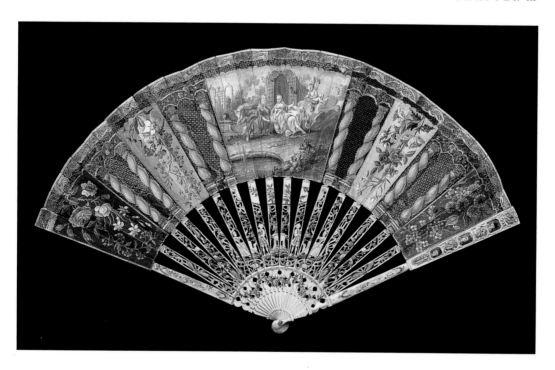

Fragonard all painted fans themselves: fan leaves thought to have been painted by Watteau are to be found in the collections of Chatsworth and the British Museum, and Fragonard is believed to have painted the fan carried by Marie Antoinette at her marriage in 1770. This, alas, has not survived, but the pretty light-coloured dresses and exuberant frivolity of paintings such as Fragonard's *The Swing* (1766) recur often in fans of the late 1760s and 1770s. This style was extremely popular at the French court, particularly with the king's mistress, Madame de Pompadour, and is often identified by the king's name, as 'Louis XV' style.

The fan in plate 14 is elaborately decorated, and the central scene illustrates the link with Boucher and Fragonard particularly well. Here two girls sit in an almost absurdly crowded garden. Even in this restricted view we can see an arched rose trellis, a lake, a wall and the stone bench on which the girls are seated. The subject is romantic as well as

Plate 14 Gouache on vellum, with insertions of cotton net. Carved ivory articulated sticks. French, 1760-70. T.19-1939.

31

Fig.4 Building a house
of cards. Detail.
Gouache on vellum.
French, *c.*1750.
T.94-1956 .

pastoral: the two girls, one with a rake in her hand, are resting from their labours in the garden, while Cupid prepares
to shoot an arrow of love towards one of them. Their round
rosy faces and brightly coloured flowing dresses are typical
of the 'Louis XV' style, and are echoed by the lavishly painted cornflowers, forget-me-nots, roses and carnations in the
side panels. Pieces of net, bordered with painted paper
columns, intersperse the panels, a popular technique in the
1760s and 1770s: at least two fans virtually identical to this
one are known. The fan also has articulated sticks: when
moved up and down a brass pin causes three ladies' heads
and then the heads of three gentlemen to be revealed,
through little windows. Such amusing tricks are common in
fans of this period, complimenting the intimate scenes of
enjoyment on the leaf.

Some fans depict scenes which seem quite out of place in
their country setting, as if their characters have been inad-
vertently plucked out of a drawing-room. In fig.4 a group sits
in a garden watching a young man building a house of cards
on a table, entertaining the mother with the baby on her lap,
while the child on the left fondles a dog. The pyramid of
cards is quite high, but the next breath of wind will presumably send the cards flying.

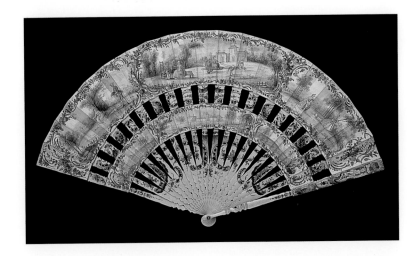

Plate 15 Cabriolet fan.
Gouache on paper.
French or English,
1755—60.
T.99-1956.

Plate 16 Harbour
scene with fishermen.
Gouache on vellum,
with carved ivory
sticks.
French, 1750-60.
T.152-1978.

In the mid-1750s a very distinctive type of fan was made,
inspired by the cabriolet carriage, invented in 1755 by Josiah
Childs. Cabriolet fans, such as plate 15, featured two or even
three separate concentric leaves on the same sticks, inspired
by the concentric wheels of the carriage. Cabriolets became
so fashionable that some single-leaf fans were painted to
look as if they had two separate leaves. The scenes were
invariably pastoral. On the top section of this leaf a small
cabriolet carriage is portrayed, flitting past a couple trying to
have a quiet drink in their garden.

Some mid-eighteenth century fans are painted with miniature classical landscapes in the manner of Claude Lorrain
(1600-82), Nicolas Poussin or, in the case of plate 16, Joseph
Vernet (1714-1789). Vernet was particularly known as a
painter of harbour scenes, and completed a set of views entitled 'Ports of France' for Louis XV. His pictures characteristi-
cally show groups of fishermen and a fishing boat in the
foreground. This fan, dating from around 1750, is faithful to
Vernet's style and is more delicately painted than many of
this period. The reverse is painted more sketchily – it was
not intended to be examined closely, unlike the front – and
shows a man chasing a dog over a bridge. The wide, closely spaced ivory sticks are typical of the 1750s, and the scene

Fig.5 A young couple in conversation. Detail. Gouache on paper with silver wash, by Johannes Sulzer (1748-94). Swiss, 1780-90. T.76-1981.

carved on the guardsticks shows a gentleman talking to a cherub who sits on a column above his head. That this scene bears no relation to that on the leaf is entirely typical; leaf and sticks were almost certainly designed by different people.

Fans like this last were undoubtedly intended to be examined and admired at parties and gatherings. They also completed a formal outfit: in Richardson's novel *Pamela*, the heroine is just about to be married and dresses herself in her best clothes: 'taking my fan, I, like a little proud hussy, looked in the glass, and thought myself a gentlewoman once more'. The 'taking' of the fan is the gesture that makes the outfit complete. But how did ladies wear or carry their fans? In many eighteenth-century portraits they let the fan hang between their fingers, closed, an indication of the importance of attractive guardsticks. In a portrait of Mrs Mary Martin by Allan Ramsay, the sitter takes her fan out of the fur muff on her lap, which may indicate where fans were stored when not in use. Very occasionally the fan is given more prominence, as in one portrait by Johann Zoffany of Queen Charlotte, wife of George III, who holds a lace fan open in front of her. Fans must have been opened and employed, not just to achieve a welcome breeze in a crowded room, but to show off their quality to envious spectators.

While most fans display loose imitations of popular contemporary paintings, a few fan painters developed their own distinctive style. Johannes Sulzer, a late eighteenth-century Swiss fan maker, was one such artist. He was also unusual in signing his fans, several of which are in museum collections. He added his address: the 'Rossignol' (nightingale), Wintherthur. His other 'signature' is the cut-out birdcage, with door open or closed, which appears on each fan. The quality of the painting detailed in fig.5 is exceptional. To fairly conventional scenes – a lady in an arbour, a man watching a bird in its cage – Sulzer adds minutely observed cameos of a spider in its web, a ladybird, and a butterfly.

No English eighteenth-century painted fans in the V&A collection can be associated with any particular painter. Sun Insurance records of the period for London do give some

Fig.6 Fan box. Robert Clarke, 26 Strand, London, *c*.1790. T.225-1959.

indication of where people went to buy fans, and list some names of fan makers. The area near the Strand seems to have been popular: in the 1720s and '30s, William Goupy was based at the sign of the 'Fann' in Surrey Street, while Joseph Jackson was at the 'Golden Fan' near Cecil Street and William West was at the 'Fan and Gown' nearby. Records also survive for fan-stick makers: when Mr Dent of Blackfriars died in December 1747, the *London Evening Post* described him as 'one of the most considerable wholesale Fan Stick Makers in England'.

Quite a lot is known about the Clarke family of fan makers. Although the only surviving fans associated with them are printed fans, their career illustrates the relative wealth and success achievable by leading fan makers. They were initially based in Ludgate Hill, from where in 1756 Robert Clarke joined the Worshipful Company of Fanmakers. Robert and John Clarke, 'Fan and Hatmaker', are then recorded with a shop at 87 Bishopsgate in 1771, moving to 26 Strand in about 1777. By 1791, Robert Clarke is working alone again, his goods are insured for a considerable sum – £800 – and he is able to name royalty among his clients. The label on a Robert Clarke fan box (fig.6) proudly proclaims the patronage of their Royal Highnesses the Duke and Duchess of Gloucester.

A significant influence on English eighteenth-century fans was the English passion for gardens. Wealthy English people recreated the wildest visions in their parks and gardens: when Richard, Lord Mornington, created a garden at Dangan in Ireland in the 1730s, he put into it 25 obelisks, a statue weighing 3 tons, a 'regular fort' with 48 cannon which saluted on family birthdays, and 3 ships on the lake, including a 20-ton man-of-war. Such excesses were much criticized in the press, but it was commonplace by the 1740s to build a temple or a gothic 'ruin', and many aristocrats created hermitages and employed a 'hermit', a contract occasionally terminated by the hermit's tendency to escape to the local tavern!

In much of this activity, gardeners were aspiring to recreate in spirit the classical landscapes of Claude Lorrain and

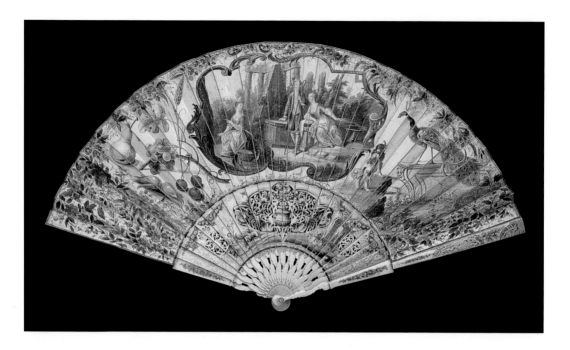

Plate 17 Gouache
on vellum.
English, 1760-70.
T.166-1920.

Nicolas Poussin, in the style known as the 'picturesque'. The formal geometric garden plans of the previous century were replaced by carefully contrived meandering paths, ponds and lakes, rocks and clumps of trees, with strategically placed urns and statues. In the *Daily Advertiser* in 1731 a Mr Langley undertook to make 'with the utmost exactness, beauty, strength and duration...all manner of curious vases, urns, pineapples, pedestals for sundials, balustrades, columns and pilasters'. Plate 32 shows a typical 'English' garden at the Villa Borghese in Rome, with a temple, surrounded by trees, at the head of a lake.

The central vignette of the fan in plate 17 might represent a fashionable English garden of the mid-eighteenth century. The main scene is overcrowded with garden buildings: a stone bench, an intentionally ruined classical colonnade, and a curious pyramid-like structure, which could be the entrance to an icehouse. The shape of the latter is influenced by Roman monuments such as the tomb of Gaius Cestius. The curious juxtaposition of buildings and people here is reminiscent of capricci, imaginary scenes painted by contemporary Italian artists. The design of this 1760s English fan is unusual, with giant still-life grapes, a peacock and a King Charles spaniel alongside the much smaller central figures.

Plate 18 probably illustrates an English fan also. Dating from the 1740s, its fresh greens recall an English summer, with vibrant red hollyhocks beside two classical urns, and a canopy of leafy branches. The accuracy with which the flowers are painted is an indication of the fan's provenance: English artists and designers were renowned for their precise interpretations of flowers and foliage. At first glance, the garden depicted on this fan appears natural, even a little overgrown, but the central figure is actually fishing in a newly built canal, her maid holding her train from behind a large, crisply cut hedge. Near the house, glimpsed in the back-

Plate 18 Watercolour on vellum. Chinese sticks, carved and pierced ivory. English, probably 1740s 1763-1869.

ground, is a boat on an artificial lake, another essential feature of the picturesque garden. Elements of the fan's decoration can be described as 'rococo', the decorative style at the height of fashion in the 1740s. The light floral border, the irregular cartouches, the sea shells and the cherubs which climb ornate curling branches up the sticks are all typical features of rococo ornament.

Both these fans show wealthy English people enjoying the pleasures of their gardens. Such gardens were deliberately designed to provide opportunities for pursuits like fishing, and featured plenty of stone benches, summer houses and arbours from which the prospects so painstakingly created could be appreciated.

Occasionally, the stone benches appear to be placed in fields, or on a hilltop, improbable rural settings for ladies in impractical silk gowns. In fact, most wealthy people confined their rambles to their gardens, but they would brave all weathers to hold 'fêtes champêtres', the popular term for outdoor meals and parties. As Horace Walpole wrote bitterly after a particularly wet summer evening at Stowe in Buckinghamshire: 'They talk of shady groves, purling streams and cooling breezes, and we get sore throats and agues attempting to realise these visions'. As ever, the English climate conspired against the realisation of these Elysian dreams.

CHAPTER IV

CHINOISERIE

Chinoiserie is clearly defined by Oliver Impey, in his publication of the same name (1977, p.10), as 'the European manifestation of mixtures of various oriental styles with which are mixed rococo, baroque, gothick or any other European style it was felt suitable'. The style is at once both light-hearted and fantastical, incorporating landscapes with improbable perspectives and flattened buildings, while scattered about are little figures dressed in what is believed to be recognisable Chinese costume.

This style of decoration is generally acknowledged to have reached its peak during the eighteenth century. The influence of chinoiserie on fans and fan design can be traced back to the beginning of the seventeenth century.

The source of inspiration for chinoiserie can be attributed to the activities of the English and the Dutch East India Companies, founded in 1600 and 1602 respectively. Both companies imported huge quantities of furniture, ceramics, textiles and fans into Europe from China and India during the seventeenth and eighteenth centuries.

The East India Company Letter Book of 1699 recorded that '20,000 fans of the finest and richest lacquer sticks' were available for purchase in Canton and Amoy. The Customs records of imported goods to London from the East Indies from Michaelmas to Christmas 1698 included the entry 'from East India, Paper Fanns 8,689 @2d pr piece'.

The Chinese export fan shown in plate 19 is an ivory brisé fan dating from between about 1720 and 1730. The decoration bears a close similarity to Chinese ceramic design. Fan painters and ceramic painters were often one and the same. Fans like this one would have been commissioned in China,

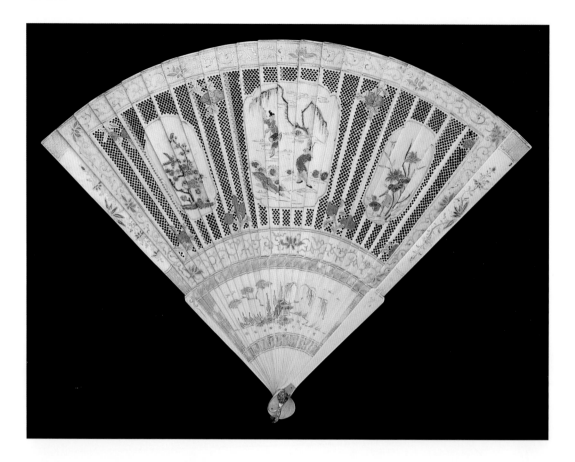

Plate 19 Pierced ivory brisé. Gouache and gilded.
Chinese export, 1720-30.
2259-1876.

probably Canton, by either the Dutch or the English East India Companies specifically for the western market. By the early eighteenth century this influx of new designs had been digested and transformed into the chinoiserie genre. Making a considerable impact on all the arts, chinoiserie inevitably became linked with another contemporary style, that of the rococo, which appeared from about the 1730s.

The chinoiserie style reached its peak in the second half of the century with the delicate and whimsical designs of Jean-Baptiste Pillement (1728-1808). A particularly fine fan (plate 20), dating from the 1760s, is composed of three charming vignettes of Chinese people. The fan's vignettes are based on designs of *c*.1758 by Pillement (fig.7), which were subsequently published in a designers' source book, *The Ladies Amusement* (1760) by Robert Sayer. Such source books provided designers with ready-made scenes, and although the designs originate from Pillement it is unlikely that the fan was actually painted by him. The leaf is of vellum and painted in watercolours with the utmost delicacy and skill. The sticks, of carved and pierced mother-of-pearl, are richly decorated in yellow gold and silver foils. The sticks are of a kind known by the French term 'battoir', because their rounded paddle shapes are reminiscent of battledore raquets or carpet beaters.

Despite the help of designers' source books, as well as a

Plate 20, overleaf. Watercolour on vellum, designs taken from Jean-Baptiste Pillement. Mother-of-pearl sticks enriched with yellow gold and silver foils. French, 1760s.
T.154-1978.

Fig 7 Designs by Jean-Baptiste Pillement from *L'Oeuvre de Jean Pillement*, première serie 1912. Engraved by P.C. Canot, 1758.

plentiful supply of original material to refer to, the depiction of Chinese features nearly always defeated European artists. Their attempts usually resulted in half-European, half-Chinese faces. Even Pillement failed; his figures are only recognizable as being Chinese by their dress, and even that is of rather doubtful authenticity.

There is a small painting in the V&A's collection – originally a fan leaf before being expanded to its current rectangular form – which shows the interior of a late eighteenth-century East India Company warehouse (fig.8). Intriguingly, the items are mainly Chinese while the scene, judging by the costumes of the people, seems to be located in either India or Indonesia and may represent one of the East India Company's Dutch warehouses. In the centre on the floor is a pile of folding fans which could be Japanese. It is quite possible that the people are imaginary as they are not wearing readily identifiable costumes. While some figures are

Fig.8 Fan leaf depicting an East India Company warehouse, extended to form a painting. Possibly Dutch, *c.*1770-1820.
P.35-1926.

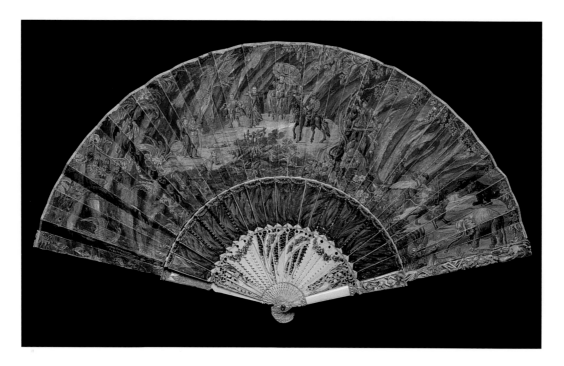

portrayed in a style that could be intended to be read as Chinese, the profile of the head of one of the central figures, for instance, looks as if it has been copied from an Indian miniature. This element of confusion is typical of European interpretation of Eastern dress and features. The furniture and ceramics are far more convincing, and could easily have been painted in a warehouse in Europe.

The element of fantasy in chinoiserie is portrayed with considerable panache on a fan of between about 1730 and 1750. It has a painted paper leaf with a completely improbable scene combining the styles of both chinoiserie and rococo (plate 21). An island floats in mid-air, while an oriental person of rank is carried on a litter and attended by a warrior on horseback. On the right-hand side is a group of rather tame-looking wild animals, whilst on the left a peacock and a heron are fighting. The whole scene is set against a rich blue background. The sticks are of carved ivory inset

Plate 21 Gouache on paper, carved and painted ivory sticks decorated with mother-of-pearl.
English or Dutch, 1730-50.
T.141-1920.

Plate 22 Gouache on paper, carved and pierced ivory sticks. English or Dutch, 1750-70. 2298-1876.

with mother-of-pearl. Painted across the reverse is an abstract scene set on a beach. On the left a large blue scallop shell and two smaller ones, in blue and mauve, are buried in the sand with just the edges showing, giving the appearance of a rippling coloured ribbon. Within each shell are sprays of flowers and clusters of exotic fruits. The centre of the leaf shows a balustrade surmounted by an urn at one end and a fantastical bird at the other. This is set next to a short flight of steps which run down to a lake containing a pair of swans. Next to the lake is a small round fountain which overflows onto the beach. The remainder of the leaf is scattered with flowers and other shells.

One of the most skilful interpretations of chinoiserie appears on a small fan of between about 1750 and 1770 (plate 22). The leaf is divided into three separate scenes which appear to be fairly accurate copies taken from Chinese paintings. Nevertheless, the artist has not quite manged to draw the faces correctly, and stylistic differences in the painting of the landscape indicate the use of European techniques. However, the carved and pierced ivory sticks are genuinely of Chinese origin. It was not uncommon for fan sticks to be imported from China, with the leaves being added later by European fan makers.

The success of the East India Companies in Europe, over a period of two hundred years, can largely be attributed to the Europeans' insatiable desire for novelty. Art objects from India, China and Japan have never lost their ability to intrigue and excite the European imagination. Chinoiserie continued as a style into the nineteenth century, but generally lacked the lightness of touch and sprightly designs of the eighteenth century. It was during this fanciful age that chinoiserie, in many branches of the arts, was employed to its best and most original effect.

CHAPTER V

COURTSHIP, WEDDINGS
AND CELEBRATIONS

Love is perhaps the most common subject of fan leaves. It appears in many guises: Venus, the goddess of love, falls in love with Adonis the hunter; lovers sit together in secluded arbours; or Cupid fires volleys of arrows at unsuspecting mortals. One late nineteenth-century fan in the V&A collection even shows a sequence of imagined scenes from the life of Cupid, setting the leaf between guardsticks shaped like an arrow shot from his bow. Such fans must often have been given to ladies by their lovers and admirers. Sadly, though, fans rarely survive with information about their original owners, so we can only speculate.

That fans were popular as tokens of admiration is evident from the letters of the eighteenth-century novelist and prolific correspondent, Horace Walpole, who often recorded gifts of fans to ladies he particularly admired. On 15 May 1742 he sends fans to Mme degli Albizzi; 'the little Albizzi' as he rather patronizingly calls her. In the same letter he refers to some 'pins and fans' he intends to present to Mme Galli. On one occasion, his friend Sir Horace Mann writes from Florence asking Walpole to purchase four fans as a wedding gift for his mistress's daughter; complicated negotiations ensue about the type of fan to buy and how to transport them to Italy.

While fashionable fans were certainly regarded as suitable wedding gifts for many ladies, and might even be carried by a bride at her marriage, fans were also made illustrating the ceremonies surrounding betrothal and marriage. For example, the 'Fête de la Rosière' was a French custom in which the most deserving young peasant girl of a village was pre-

sented with a garland of roses and a small dowry by the local landowner, thereby increasing her chances of making a respectable marriage. Many French fans, including three preserved at the Musée de la Mode et du Costume in Paris, illustrate versions of this ceremony. These might be considered to be marriage fans, but their charming, idealized visions of rural life would have appealed to most fashionable Frenchwomen in the mid-eighteenth century.

Some commemorative marriage fans were the equivalent of the modern royal souvenir. One example in the V&A collection has a painted leaf commemorating the marriage of the Dauphin, the future Louis XVI, to the Austrian princess Marie Antoinette, in 1770. The fans illustrated in this chapter, however, are on the whole anonymous. Dating from the eighteenth century, they all incorporate similar themes and imagery, which link the fans with betrothals or marriages.

Plate 23 Gouache on thick vellum. North European, *c.* 1700. T.246-1990.

Probably the earliest marriage fan in the collection dates from *c.*1700 (plate 23), and has an unusual and particularly enigmatic design. Its leaf is of especially thick vellum, painted in gouache with a symmetrical design of baroque swags and flowers, some resembling the Tudor rose motif. In the centre are two hearts, surmounted by a golden crown. The crown probably represents a couple's 'coronation' as man and wife, rather than any royal connections. The design of the leaf is unique, quite unlike that of any other surviving fans of this date, suggesting that it was dreamed up by a provincial fan maker for a bride who was perhaps not familiar with current fashions.

The use of paired hearts to symbolize a marrying couple recurs later in the century. One fan at the Fitzwilliam Museum, Cambridge (M.217-1985) is painted with the rococo floral sprays and bows typical of fans of the 1770s. Its central cartouche shows a group of musical instruments – music has always been a symbol of romantic love – while on either side are hearts decorated with swags and golden spangles, but still recognizable.

Most eighteenth-century marriage fans, however, are less symbolic; the marriage ceremony itself is a common theme. It is usually the drawing up of the formal contract by lawyers that is illustrated, rather than a church service. This indicates the importance given to the exchange of land or money associated with marriage for prosperous families. In fact, studies of marriage traditions in England show that many unions were contracted without a church ceremony taking place, until the Marriage Act of 1753 finally made this illegal.

One of the best examples of a marriage contract fan in the V&A collection dates from 1760-70 (plate 24), and shows the wedding apparently taking place in the kitchen of a prosperous provincial household. The style of the fan suggests it is of English or Dutch origin. The bride and groom rather self-consciously hold hands in front of the two lawyers, who are seated at a table with the marriage documents in front of them, discussing some aspect of the procedure. A group of witnesses sits behind the couple, whispering and chatting.

To the right, another woman is laying the table. The flagons of wine and barrels of ale lying nearby suggest that a hearty wedding feast is in preparation.

The floorboards in this kitchen are bare, and the cupboards are of solid provincial design. The clothes, too, are simple and practical, suggesting that these are the servants in a large household, celebrating a 'below stairs' marriage. This is an expensive fan, however. The sticks, alternately of ivory and tortoiseshell, are of high quality, as is the painting on the leaf. It is tempting to assume that it is of provincial manufacture, the scene painted by someone unfamiliar with fashionable interiors or dress. The more likely explanation is that the scene is taken from a contemporary play or novel, and was illustrated on the fan to provide a talking point at a society gathering.

The other aspect of the marriage ceremony to attract fan painters was the wedding banquet. The fan illustrated in plate 25 is a particularly charming example. Probably from

Plate 24 Gouache on vellum, alternate ivory and tortoiseshell sticks. English or Dutch, 1760-70.

T.385-1910.

Plate 25 Gouache on vellum, ivory sticks. Dutch or German, 1740-50. 2211-1876.

Fig.9 Detail from reverse of plate 25.

Holland or Germany, and dating from 1740-50, it shows several bucolic scenes from a feast. The couple in the centre of the fan, in richer clothes than the guests, appear to be recently married. The scene on the back of the fan (fig.9), showing a couple inspecting the building of a new house, seems to confirm that the front depicts a marriage feast. The artist has included many humorous touches, such as the man stealing a bottle of wine from the crate on the left.

The most spectacular marriage fan in the Museum's collection is shown in plate 26. It portrays the altar of Hymen, goddess of marriage, in a particularly flamboyant manner. As the bride and groom approach, clouds billow around the altar, a rainbow springs out of the earth, and gods and angels hold aloft elaborately framed portraits of the happy couple. These were undoubtedly intended to be recognizable likenesses of a young couple of the French or Prussian nobility, although their identities are not now known. Exquisite still-life vignettes at either side contribute to the air of wealth and

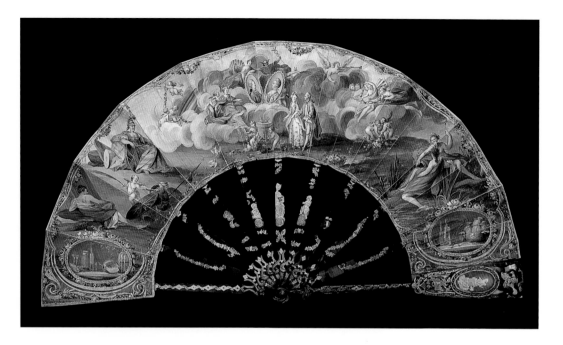

Plate 26 Gouache on vellum, tortoiseshell sticks with applied silver gilt and pink foil decoration.
French, 1750-60.
T.98-1956.

Plate 27 Banquet scene. Detail. Reverse of plate 26.

opulence which pervades the scene, and it is no surprise to find that on the reverse (plate 27) a magnificent banquet, using fine silver and glass and featuring liberal supplies of wine, awaits the guests, albeit placed improbably in a pas-

Plate 28 Gouache on silk, embroidered with sequins. Tortoiseshell sticks.
French, 1780-90.
97-1864.

toral landscape. This style of fan, with widely spaced sticks and a half-circle narrow leaf, is typical of the 1750s. The sticks are of tortoiseshell, with silver gilt and pink foil vignettes on the guardsticks.

Marriage between couples of disparate social backgrounds was a popular subject with eighteenth-century writers, especially the authors of comedies for the London theatres. Garrick's play *The Clandestine Marriage* (1766) tells the story of the unsuitable love of Fanny, the daughter of a wealthy merchant, for Lovewell, her father's clerk. She spurns the wealthy Sir John Melvil, and true love wins in the end. Alongside the endless stream of such plays, most fashionable ladies would also have a copy of Samuel Richardson's celebrated novel *Pamela*, which describes the courtship of a lady's maid by her widowed master.

The scene in the central vignette of the French fan in plate 28 clearly suggests a storyline concerning the odd assortment of characters illustrated: the prim young girl sitting at the edge of her seat, the large, rather slovenly woman next to her, and the gentlemen shaking hands. Could the seated gentleman be a nobleman congratulating the young man in blue on his marriage to the girl? Whatever the story, this is clearly not meant to be an accurate representation of a fashionable marriage. The vignettes on either side, however, show ladies posing in a country park in the elegant, fashionable dress of the early 1780s. These scenes may be derived from the popular engravings of contemporary costume by Jean Michel Moreau, known as Moreau le Jeune. The presentation of this fan, painted on a silk leaf with sequins attached and with highly decorated horn sticks, is also typical of the 1780s, when sequinned silk leaves became generally popular.

Marriage was a rich subject for satire as well as celebration in the eighteenth century. So often used as a means to acquire wealth, to get rid of a troublesome daughter, or to gain status and power, it inspired a certain degree of cynicism. A scene from William Hogarth's famous series 'Marriage à la Mode', engraved in 1745, shows two fathers and lawyers arguing over the financial details of a settlement for an unloving couple (fig.10). Infidelity was widely accepted in fashionable society; the Fitzwilliam Museum has a painted French fan of the 1770s (M.300-1985) entitled 'Le Mari Moderne' ('The Modern Husband'), which shows a lady flirting with two ardent gentlemen, one a cleric, while her husband stands patiently by with a string attached to his nose, held at the other end by his wife.

A printed fan leaf at the V&A is entitled 'The Good-for-Nothing Swain' (fig.11). Printed by J. Read of 133 Pall Mall in 1796, its three vignettes illustrate 'The Vow of Constancy', 'The Hour of Infidelity', and 'Cupid's Farewell'. Similarly, a printed leaf in the Schrieber Collection at the British Museum entitled 'Before and After Marriage' compares the kindness of the lover with the 'indifference' of the husband. The lurid romantic life of the Prince of Wales, later Prince Regent, and in particular his relationship with Mrs Fitzherbert, inspired many satirists of the time, and certainly fuelled the produc-

Fig.10 'The Marriage Contract'. Detail. Plate 1 from 'Marriage à la Mode', contemporary print after William Hogarth. English, 1745. Dyce 2750.

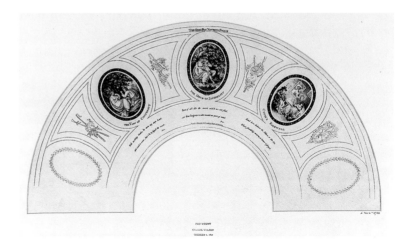

tion of printed fans that portrayed marriage in less than
favourable terms.

It would be a pity, however, to close this discussion of
marriage fans on a cynical note. Fans do survive that celebrate marital success and happiness. At the Musée de la
Mode et du Costume in Paris a delightful painted fan of
*c.*1775 shows a man with his wife and two young children.
The two vignettes at either side show ships at sea. It seems
probable that the fan commemorates a naval officer's home-
coming to his wife and young family.

However, it is in the V&A's collection of nineteenth-century fans that the most personal and touching celebration of
a marriage is to be found. It is the fan painted by Sir Matthew
Digby-Wyatt as a tribute to his wife, Mary, in 1869 (see plate
1). Part of its inscription reads: 'For Love is Heaven, and
Heaven is Love' and it is signed 'M Digby Wyatt To Mary
Wyatt 1869'.

CHAPTER VI

SOUVENIRS FROM ABROAD

Rome, Tuesday 15th May, 1714. 'Paid for a fan painted by Mr Winter a German 1.50.0' [Roman crowns].

Thus reads a single tantalizing reference to an interest in fans from the account book of the Grand Tour of Thomas Coke, later 1st Earl of Leicester. One of the most passionate art collectors ever to undertake a Grand Tour in Europe, Coke built a magnificent house at Holkham in Norfolk on his return, where many of his treasures are still displayed. We do not know whether the fan he refers to was given to a lady in England or kept for visitors to admire in the Library at Holkham, as no fans of this period have survived in the collection of that great house.

Coke's fan probably illustrated a theme from classical mythology or the Old Testament. Fans depicting specifically Italian scenes do not generally appear until the mid-eighteenth century. One of the earliest dated examples is by Joseph Goupy, an English watercolourist and draughtsman known for his landscapes in the style of Salvator Rosa. Dated 1738, the fan shows the Arch of Constantine, the Arch of Titus, and the Forum in Rome. Goupy worked in London but, like most artists of his time, must have undertaken an artistic 'pilgrimage' to Rome. Other fans showing Roman scenes attributed to him also survive: all are decidedly Italian in quality and style.

While Thomas Coke's fan and the fans of Roman scenes by Joseph Goupy would all commonly be described as Italian, Italy was a melting-pot of artistic talent from all over Europe. Many artists painted fans, and not always by choice.

In 1787, Aloys Hirt listed 400 Italian and 163 foreign artists living in Rome. Those who struggled to find work often painted and sold fans to make ends meet.

In the 1770s the increasing vogue for a style of ornament influenced by recent excavations at Herculaneum and Pompeii was taken up by fan painters. At the same time, the term 'tourist' began to be used to describe the young, often wealthy, Englishmen travelling on the 'Grand Tour' route to Italy. Not all were enthusiastic about history or the arts, but they all wanted souvenirs of their trip. It is hardly surprising, then, that fans showing the great sights of Italy, often decorated with fashionable neo-classical ornament, should have been popular. The V&A has collected a number of these 'tourist' fans, which are the subject of this chapter.

The wealthiest 'tourists' travelled with several coaches and a huge entourage of tutors, companions, and hangers-on. The 3rd Earl of Burlington set out from Burlington House in London on 17 May 1714 with two coaches, each drawn by a pair of horses, and a huge retinue of outriders and liveried servants. He was accompanied by Joseph Goupy's uncle, Lewis Goupy, an artist and fan painter, whose job it was to record memorable images of their travels.

Although there was no standard route, Paris was an important destination, and other cities in France were visited, as well as in the Netherlands, Germany, Austria, and Switzerland. Along the way, travellers found hotels, taverns, and a variety of (often murky) entertainments, as well as a bewildering array of cultural pleasures. Almost without exception, however, the highlight of any Grand Tour was the journey through Italy, and it is with the Italian leg of the trip that we must now concern ourselves.

For most, Venice was the first major stop. While its architecture was little appreciated – the historian Gibbon condemned its 'ill-built houses...and stinking ditches' – travellers revelled in the casinos, coffee houses and brothels, in the endless masques and parties, and above all in the city's February Carnival. It is telling that Thomas Coke bought no paintings at all in Venice, and appears to have visited neither churches nor monuments, but that his servant's account book refers to countless purchases of tickets for the opera, and for masked balls.

Carnival was the highlight of the year: crowded, chaotic and flamboyant, but endlessly fascinating. One guide book of 1739, F.M. Misson's *New Voyage to Italy*, describes the 'strangers and courtezans [who] come in shoals from all parts of Europe', and the 'general Motion and Confusion, as if all the World were turned Fools, all in an Instant'. The great Masquerade in St Mark's Square is given special attention: 'you must be able to maintain the character of the person whose dress you have just taken...When the Harlequins meet they jeer one another, and act a thousand fooleries. The Doctors dispute, the Bullies...swagger'.

The fan illustrated in plate 29, usually entitled 'The Triumph of Harlequin', evokes the spirit of Carnival. Its subject appears to derive from a design for a fan painted by Felicita Tibaldi Subleyras, a Roman artist, for Elisabeth Farnese, the Italian-born Queen of Spain. This unusual design, in oil on canvas, is in the Musée des Augustins in Toulouse, and seems to have provided the source for at least one other surviving fan as well as the V&A's example.

This is interesting evidence of the duplication of a high-quality fan. We cannot know who purchased each example, but it is hardly surprising that it should have been in demand. Very finely painted in watercolour on kid, using a curious stippled technique, the sheer quality of this fan would satisfy a collector of art, and the design, which Subleyras based in turn on a painting by Hubert Robert (1733-1808), is in the same tradition as the fashionable Carnival scenes of Pietro Longhi (1702-85). The scene on the reverse is equally fine, showing a tall ship anchored near a dark cliff. For a tourist, however, the dark masks and vibrantly coloured clothes, together with the gleam of the mother-of-pearl sticks, would be a fitting memento of the excitement of the Venice Carnival.

Although few fans of this period are unquestionably of Venetian origin, several examples showing the Rialto bridge,

Plate 29, overleaf. 'The Triumph of Harlequin', after a design by Felicita Tibaldi Subleyras. Italian, *c.*1750. T.153-1920.

59

Plate 30 Scenes of Rome.
Gouache on vellum.
Italian, 1790-1800.
2171-1876.

a universally admired feature of the city, have survived.
Copies of scenes by the most popular Venetian painter of the
day, Canaletto (1697-1758), can also be found on fans. One
example, showing St Mark's Square, fills the whole fan leaf
with the image, leaving no room for any additional ornament, and giving the owner an accurate and detailed, if
somewhat reduced, 'Canaletto' to take home.

After admiring the treasures in the renowned Uffizi
Galleries in Florence, and perhaps visiting Bologna, a centre
of painting and fan making, the 'tourists' journeyed on
through Tuscany and Umbria to Rome, for most the glorious
highlight of their travels. For Thomas Coke, this was the
moment to start buying paintings. On his two trips to Rome,
in 1714 and 1716, he collected obsessively. His servant
Edward Jarrett records the transactions, sometimes with a
note of weariness, as on 1 May 1714: 'paid for pictures 3
times today'. When the Earl of Burlington finally returned
home he is said to have travelled with 878 pieces of luggage,
most of them containing works of art.

A valuable addition to the library, or to the print collection, was a fine example of miniature painting such as a fan.
Fans bought for this purpose were purchased as flat fan

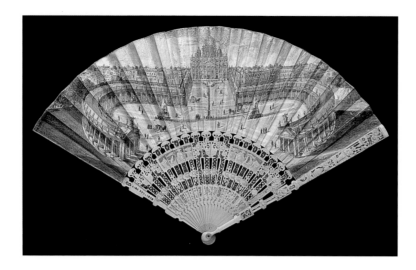

Plate 31 St Peter's, Rome. Gouache on vellum. Italian, 1770s. Circ.377-1959.

leaves, many examples of which survive in museum collections. However, some of the fan leaves which were bought unmounted were later mounted up in London, taking account of the latest fashion for sticks.

The most common type of tourist fan illustrated views of well-known 'sights'. One favourite in Rome was the Piazza del Popolo, a fine square situated at the entry to Rome from the north and the first memorable view of the city, only slightly spoiled by the goats wandering around and the hovels of washerwomen and prostitutes nearby. The fan in plate 30 shows the Piazza del Popolo in the centre, a view perhaps influenced by a fine contemporary painting of the square by Giovanni Pannini (*c.*1692-1765/8). To the right of this image, and set within fashionable neo-classical borders, is the Colosseum, drawn and painted by Ducros, Piranesi, and every other artist in the city. The tomb of Cecilia Metella, shown on the left, is perhaps less well known now. According to Misson's *New Voyage to Italy*, its chief claim to fame was 'that curious echo' which was said to repeat back 'a whole Hexameter verse'. Somewhat unsurprisingly, however, when Misson visited, the echo seemed to have disappeared.

In plate 31, dating from the 1770s, the image of St Peter's fills the fan leaf. The perspective has been slightly distorted to give a sense of the great vista leading to the church, although Lady Anne Miller, visiting Rome in 1770, thought prints on a small scale 'give but a very faint idea of the magnificent original'. She was not all that impressed by the façade anyway, feeling that with 'so many ornaments, such twisting and turning...one fine thing hides another'. Nevertheless, the shape of the building, with its great looping colonnades, fits the fan leaf perfectly. The ivory sticks are carved with fashionable chinoiserie scenes.

A more rural scene, the English Garden at the Villa Borghese in Rome, is illustrated in plate 32. Designed by Jacob More, this garden was completed in 1787, and incorporated a Temple of Aesculapius by Mario Asprucci, which appears in the central vignette. This Italian fan, probably made for an English tourist and showing an English landscape garden designed for Italy by an Englishman, illustrates well the complex cultural interchange between England and Italy in the eighteenth century. The fan's other distinctive feature is the sticks, alternately of ivory and wood.

Some of the finest surviving tourist fans look to paintings, rather than architecture, for their subjects. The central vignette of the fan illustrated in plate 33 comes from a fresco by Raphael, at the Villa Farnesina on the Palatine Hill, of Cupid ordering Venus to punish Psyche for her vanity. Lady Anne Miller, describing the arcade in which Raphael's series of paintings could be viewed, noted that recent restoration had 'heightened some of the backgrounds with a kind of blue colouring'. This surely explains the vibrant blue behind the figures of Cupid and Venus on this fan. The miniature copy is certainly worthy of the original, painted on fine, crisp vellum with considerable technical mastery.

The two subsidiary vignettes are taken almost directly from images of dancing nymphs found on wall paintings at Herculaneum, the Roman town buried under volcanic ash near Naples and discovered in 1738. These and many others were illustrated in a series of volumes entitled *Le Antichità di Ercolano esposti*, published from 1757 onwards but not

Plate 33 Central vignette after Raphael. Gouache on fine vellum.
Italian, 1790-1800.
161-1899.

Fig.12 'The Cupid Seller', from *Le Antichità di Ercolano esposti*, vol VII p.41 (1762). National Art Library.

Fig.13 Furnishing panel, woven silk. French, 1800-15. 1274B-1871.

generally available until the 1770s. Several images found their way onto fans, including 'The Cupid Seller' (fig.12) which appears on a fan in the Badisches Landesmuseum, Germany. Versions of the nymphs not only appeared on fans, but also on wall paintings, ceramics and textiles (fig.13). The thrill of visiting Herculaneum and Pompeii certainly drew visitors by the 1770s. Lady Miller writes of being taken down to the subterranean theatre of Herculaneum by the light of flickering torches: she was impressed, but made a hasty exit from 'so unwholesome an air'.

The buried cities were a new attraction, but Mount Vesuvius, the brooding volcano whose lava had destroyed them, had been visited and often climbed by almost every traveller in the eighteenth century. Sir William Hamilton, British Envoy in Naples from 1764, was obsessed with the volcano, which he climbed 25 times in four years, and his book about Vesuvius, *Campi Phlegraei; Observations on the volcano of the two Sicilies*, was published in 1776. Its dramatic illustrations by Pietro Fabris were doubtless known to fan painters. A series of eruptions in the 1770s and 1780s made this a topical subject for tourist fans; an example is shown in plate 34.

This fan is distinguished by remarkable ivory sticks with a delicately carved classical scene and, on the guardsticks, urns picked out in blue enamel and gold paint. The detail, fine carving and gold highlights suggest that these sticks were particularly costly, although the fan leaf is not exceptional. The sticks may have been added back in England to make the fan a more desirable present.

It is worth noting the 'postcard' style of this fan; each vignette is set in a trompe l'oeil frame, which is annotated in the corner with a short description of the scene. The central vignette shows the arches of the Bridge of Caligula, then still visible above the sea. The bridge, which crossed part of the Bay of Naples at Pozzuoli, was traditionally said to have been built by that most notorious of emperors. Misson's *New Voyage to Italy* described it as 'so bold a Piece of Work, that it...might be ranked among the greatest Prodigies'. The 'Grotta da Cane', depicted on the left of the fan, is at

Plate 34 Views of the bay of Naples. Carved ivory sticks, with urns picked out in blue enamel and gold paint. Italian, 1779-90.
T.88-1956.

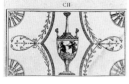

Fig.14 Plates C to CII from *Designs and Ornaments*, Michel Angelo Pergolesi, 1777-82. Engraved by Bartolozzi after Cipriani.
E.1513-1907.

Plate 35 Etched and hand-coloured in gouache on vellum. Printed by Francesco Bartolozzi (1725-1815). English, 1779.
T.7-1935.

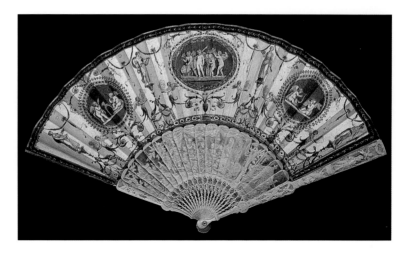

Posilipo, also on the Bay of Naples, an area much visited by tourists eager to explore the caves there. The 'Grotta da Cane' was so called because it was said to have a terrible effect on dogs, which on entering the cave succumbed to convulsions and had to be dragged out.

By 1800, the French revolutionary wars had put a temporary stop to the Grand Tour, but travellers recently returned from Italy were already at the forefront of the revolution in taste in the decorative arts. Some had persuaded leading Italian artists to come with them to England. Giovanni Battista Cipriani and Michelangelo Pergolesi, both protégés of the architects Robert and James Adam, collaborated on important design books. The fan in plate 35, printed by one of the best-known engravers in London, the Italian Francesco Bartolozzi, uses vignettes imitating sculpture 'reliefs' taken directly from Pergolesi's book *Designs and Ornaments* (fig.14). The other motifs on this fan, such as the mermaid-like cherubs, are taken directly from Herculaneum wall paintings illustrated in *Le Antichità di Ercolano esposti*. The fan illustrates perfectly the new synthesis of sources taken directly from the classical world with the work of contemporary neo-classical designers.

CHAPTER VII

BRITISH COMMEMORATIVE FANS

Whether intended as long-term souvenirs or ephemeral curiosities, eighteenth- and nineteenth-century fans were often made to commemorate important events. They marked the births, marriages and deaths of well-known people, royal occasions or major social events, as well as significant moments in political or military history. They can offer fascinating insights into the attitudes and loyalties of their time.

The V&A collection of commemorative fans is by no means comprehensive, and relates mainly to eighteenth-century Britain. The emphasis is also on higher quality decorative examples, whereas some of the most intriguing and informative designs were used for cheaply printed paper fans with plain wooden sticks. These were produced in large numbers from the early eighteenth century, but as they were not considered valuable and often disintegrated quickly, they are now extremely rare. Since 1959, however, the V&A has managed to acquire a handful of interesting examples, two of which feature in this chapter.

The European wars waged between 1700 and 1740 inspired many printed fans, their heroes and battles perhaps encouraging an increase in the production of such ephemera. Certainly, many fans survive, both English and French, commemorating the victories of the English general John Churchill, later 1st Duke of Marlborough (1650-1722) at Blenheim and Ramillies. It is impossible to establish with certainty when and by how much the production of fans was stepped up, but a number of examples survive from the 1730s. In the same decade *The Craftsman*, a political journal, contains many references to new fan designs being issued.

Plate 36 illustrates a politically inspired fan of which several copies are now known, in both English and French versions, though the misspellings and inaccuracies in the French text betray its English provenance. Its English title is 'A New game of Piquet now in play among the different nations of Europe', and it is, in fact, an unusual English commentary on the War of the Polish Succession (1733-8), in which Britain effectively played no part. The pretext for the war was a dispute over who should be elected King of Poland on the death of Augustus II in February 1733. When the final peace treaty was signed on 2 May 1738, France had succeeded in her real goal – the annexation of Lorraine – having conceded the Polish throne to the Elector of Saxony, the candidate preferred by Russia and Austria.

The fan dates from late in 1734, by which time France had effectively won, but nothing was finally decided. The seven countries participating in the war – Sardinia, France, Spain, the Austrian Empire, Saxony, Russia and Poland – are all playing piquet. The comments of the players are cleverly matched to events. To quote from the English version, France is saying, 'I make the hand and play first'. Britannia's attitude is particularly well expressed: 'I'm preparing though I don't play, but if I am nettled I'll take up the cards'. The British Prime Minister, Robert Walpole, refused to participate in the war, despite the king's wishes. At the end of 1734, he is said to have announced proudly that 'there are fifty thousand men slain this year in Europe, and not one Englishman'. The man in black next to Britannia holds a scroll saying, 'Tis not the interest of the nation to play without advantage. In time, Commerce might pay the Cards'. This echoes Walpole's view that England stood to lose too much ground economically if she joined in, and suggests that the fan might have been produced by his supporters. It is intriguing that French versions should exist, however, and that so many fans of this design – in all about ten – have survived. Perhaps one clue to their preservation is the design's relatively fine printing and delicate hand-colouring. The sticks too are of bone, not wood, the cheaper alternative.

Britain could not always choose to avoid military conflict. A constant threat in the first half of the eighteenth century was the possibility of invasion in support of the Catholic Jacobite claimants to the British throne. George I was the first Hanoverian monarch of England, succeeding Queen Anne in 1714. Supporters of the Old Pretender, the exiled son of James II who had fled England in the Glorious Revolution of 1688, were determined to ensure his succession. Consequently, parliament and George I were in constant fear of a military attack from the Catholic countries of Europe. To support the Jacobite cause was treason against the reigning monarch, and few active Jacobites dared to have objects in their possession which declared their allegiance. Genuine eighteenth-century Jacobite objects therefore, including fans, are rare. The V&A, however, has two examples which between them help to tell the story of the Jacobite rebellions.

The fan in plate 37 dates from the 1720s. The motifs on it are all significant, and are, from left to right: Charles II hiding in an oak tree at Boscobel House after his defeat by Cromwell at the Battle of Worcester; Queen Anne ascending to heaven; a lady mourning the loss of the crown; two putti holding the Stuart arms; and the white rose associated with the Stuart claim to the English throne, with its two buds representing the Old Pretender's two sons.

Queen Anne died on 1 August 1714. As the half-sister of

the Old Pretender, and daughter of James II, she was the last Stuart monarch of England. On her death, George, Elector of Hanover, a mere great-grandson of James I, was asked to become king. The Jacobites, having mistakenly believed that James III would be chosen, were taken by surprise, but a year later sent a force to Scotland to initiate a rebellion. The somewhat farcical Battle of Sheriffmuir ensued, from which both sides fled, and the Jacobites withdrew, unable to secure enough assistance. This fan laments both the death of the Stuart queen and the end of the first Jacobite rebellion. It is exquisitely painted in gouache on vellum, and the simple sticks are of high quality – it is certainly a fan its owners intended to keep as a symbol of their support for the Jacobite cause.

In August 1745, Prince Charles Edward Stuart landed in Scotland, with a shipload of troops, to fight for his father's claim to the throne. After an early success at the battle of

Prestonpans, the Jacobite campaign grew increasingly desperate, and was finally stamped out at the battle of Culloden on 16 April 1746. Fig.15 shows an English fan celebrating the victory of the Duke of Cumberland over the fleeing rebels at Culloden. The Duke is surrounded by Scottish lords kneeling in surrender. Although the true number of participants and casualties is hard to establish from the propaganda of both sides, it is certain that the rebels were slaughtered in vast numbers, while the king's troops suffered few casualties. One English commander wrote: 'The insignificance of our loss, considering the fury and despair with which they attacked, is hardly credible'. On this printed fan, the English troops are seen shooting the fleeing rebels. There were apparently piles of bodies littering the road all the way back into Inverness, and the subsequent atrocities earned Cumberland the nickname 'The Butcher', which he never shed.

Among surviving mourning fans the most common are those relating to royal deaths, when an appropriate fan had to be carried at court. Royal mourning fans often show Britannia weeping. In fig.16 Britannia sits on the right while the allegorical figure associated with Hanover, carrying a shield with the Hanoverian horse emblem, is also mourning

Fig.15 Surrender of the Jacobite leaders to the Duke of Cumberland after the Battle of Culloden. Detail. Printed on paper and hand-coloured. British, *c.*1746.
T.205-1959.

Fig.16 Mourning fan, probably marking the death of Frederick, Prince of Wales in 1751. Gouache on vellum. English, *c.*1751.
T.202-1959.

her loss. The fan probably commemorates the death of Frederick, Prince of Wales, heir to George II, in 1751. Like most mourning fans of this period it is painted in black gouache on a white vellum ground. The sticks are typically simple, of carved ivory. However, the story behind this fan is not as straightforward as it might appear. The death of Frederick Prince of Wales was a politically charged event. His stormy relationship with his father, George II, had broken down completely by 1737 and the young prince had then set up his own court. It was common knowledge that the king loathed Frederick and much preferred his second son, the Duke of Cumberland, hero of Culloden. Frederick had been an enthusiastic patron of the arts and supporter of young artists. He is also known to have employed the noted fan painter Joseph Goupy to give drawing lessons to his son George, the future king.

Frederick's son succeeded to the throne as George III in 1760. A shy man, he preferred the company of his beloved wife Charlotte and their children in the peaceful surroundings of Windsor to the public show of court life. However, recognizing the importance of developing a positive image for the royal family, he was careful to appear at public occasions with his children and actively encouraged painters to record such events. The fan in plate 38, of which several examples survive, is a version of a print by Pietro Antonio Martini, taken from a painting of the king and queen and their thirteen children at the Royal Academy Exhibition of 1788. George III was particularly proud of the achievements of the Academy, which he founded in 1768. The room shown is in Somerset House, now home to the Courtauld Institute of Art, where the Academy was initially based.

This is a high-quality printed fan made by Mr Poggi of St George's Row, Hyde Park, a highly regarded fan maker. Fanny Burney, in her diary for March 1781, describes a visit to his shop with Sir Joshua Reynolds. The fans 'are indeed more beautiful than can be imagined. One was bespoke by the Duchess of Devonshire for a present, that was to cost £30'. Here Poggi has employed a 1790 version of the origi-

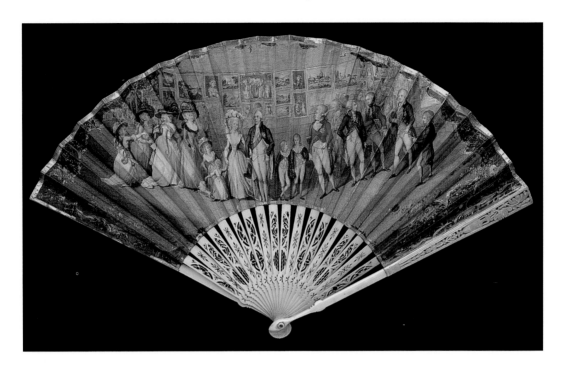

nal Martini print, which omits Sir Joshua Reynolds and other leading Academy figures, and adds three of the king's youngest children to create a complete family portrait.

During 1788 George III became ill, tormented by a disease now thought to be porphyria, but at the time believed to be madness. He was subjected to painful and humiliating treatments before his recovery in February 1789. His illness had provoked a constitutional crisis, and the relief and joy at his recovery was expressed in exuberant celebrations. On 10 March 1789 London was ablaze with fireworks. One commentator thought it 'the most brilliant, as well as the most universal exhibition of national loyalty and joy ever witnessed in England'. This fan is surely an affirmation of the king's return to normal life.

The fan in plate 39 is also associated with the celebrations for the king's recovery. Inscribed 'Health is restored to ONE and happiness to Millions', and dated 1789, it is almost iden-

Plate 38 Hand-coloured printed fan, based on the painting of George III and his family at the Royal Academy Exhibition of 1788, by Johann Heinrich Ramberg (1763-1840). English, *c*.1790.
T.56-1933.

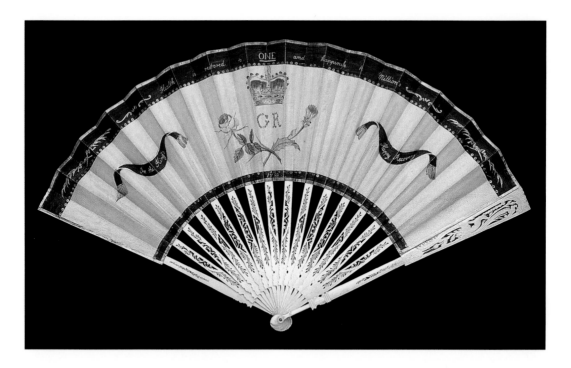

Plate 39 The recovery of George III from illness.
English, dated 1789.
T.203-1959.

tical to several other surviving fans, which may have been carried at the ball to celebrate the end of the king's illness. The rose and thistle represent the Union of Scotland and England, achieved by Act of Parliament in 1707.

George III died, blind and senile, in 1820. His reign had been a relatively peaceful one, but it is worth noting that the turbulent events on the Continent produced very different commemorative fans. The Musée Carnavalet, Paris, has a particularly good collection of fans commemorating events of the French Revolution. Every development, some now too obscure to recognize, was subject to propaganda, and huge numbers of printed fans were produced. A famous example is the 'Assignat' fan, which showed a trompe l'oeil image of a pile of bills of exchange, all worthless as inflation soared and governments changed overnight.

The V&A has few nineteenth-century commemorative fans, and no royal fans from this period to compare with the

Georgian examples described above. However, huge numbers of such fans, both printed and painted, continued to be produced. For example, the Musée Carnavalet, Paris, has a particularly charming painted fan by the French fan makers Duvelleroy, showing Queen Victoria with Prince Albert and their children at Osborne House, their family home on the Isle of Wight. Based on a painting by Franz Zaver Winterhalter (1805-73), and with images of Windsor Castle and the Isle of Wight on the reverse, the fan presents an image of quiet family happiness reminiscent of the view of King George III and his family seen in plate 38.

While the V&A collection includes no nineteenth-century fans commemorating military events, there are examples marking contests of another sort, the displays of industrial power and prestige that formed the great international exhibitions. Fans, varying in price and quality, were sold as souvenirs at all the major exhibitions and large numbers survive

Plate 40 Souvenir of the Great Exhibition, 1851. Printed and hand-coloured on paper. English, 1851. T.290-1971.

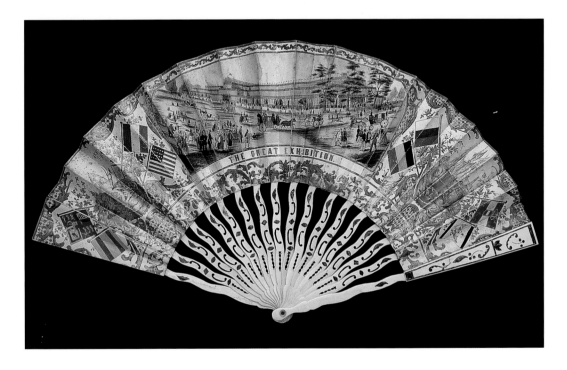

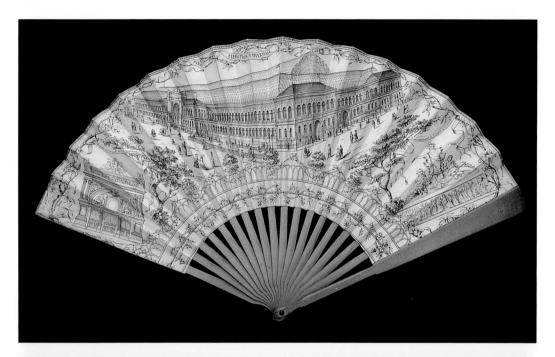

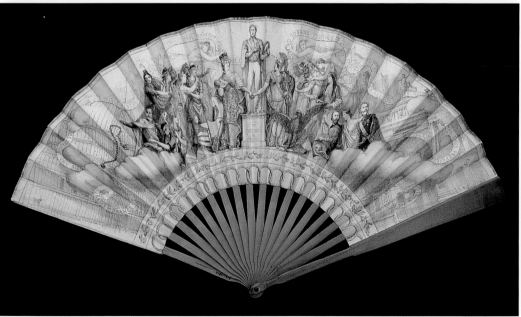

today. Plate 40 shows a fan commemorating the first truly international exhibition, held in 1851 in Joseph Paxton's specially built glasshouse – the 'Crystal Palace' – in Hyde Park. Probably one of the more expensive fans on sale, its finely printed leaf shows the Crystal Palace along with the flags of participating nations; the sticks are of bone rather than wood. Marvelling at the huge Crystal Palace, people flocked in unprecedented numbers to see the displays of raw materials and manufactures from all over the world. On 7 June Charlotte Brontë wrote to her father, 'It may be called a bazaar or fair, but it is such a bazaar or fair as Eastern genii might have created. It seems as if only magic could have gathered this mass of wealth from all the ends of the earth'. Fine art was excluded, as the exhibition was intended to celebrate advances in industry, but fans from France and Spain, including examples by Duvelleroy, were on show.

The French held an Exposition Universelle in Paris in 1855. Based in the newly built 'Palais de l'Industrie' on the Champs-Elysées, it was larger than the 1851 Great Exhibition but was not as successful, and made a huge financial loss. The Palais de l'Industrie is illustrated in plate 41, the reverse of a Duvelleroy fan dated 1862. It is perhaps indicative of French-English rivalry that a fan evidently printed for visitors to the forthcoming 1862 Exhibition in London should illustrate the site of the Paris Exhibition. However, its front (plate 42) commemorates the recent death of Prince Albert, one of the main forces behind the triumphant exhibition of 1851, and shows his statue with Queen Victoria standing below. The inscription marks his achievement, which brought 'together in close relation the people of all countries and has thus prepared the happiness of the world'.

Those who love the buildings and collections at the Victoria and Albert Museum must echo this tribute. The profits from the Great Exhibition enabled the purchase of the land in South Kensington on which the South Kensington Museum, now the Victoria and Albert Museum, was built.

Plate 41 Reverse of plate 42.

Plate 42 Duvelleroy fan, commemorating the death of Prince Albert in 1861 and the 1862 International Exhibition. French, 1862. T.185-1979.

CHAPTER VIII

BRISÉ FANS

By the end of the seventeenth and the beginning of the next century, the European brisé fan had developed from the rather large and clumsy construction found in the early seventeenth century to a small, delicate ivory fan painted in gouache or pure watercolour. The small imported ivory brisé fans from China with over twenty sticks were available at the start of the seventeenth century through the merchants of the English and Dutch East India Companies. At the height of their popularity, folding fans or just the fan sticks, mainly of Chinese design, were imported by the thousand into Europe. The ivory brisé fan of this period was small, approximately nine inches long (23cm) with between 20 and 30 flat, tapered sticks which overlapped each other by about an eighth of an inch (4mm). At the top of the leaf the sticks were held in place by fine silk cords threaded through small holes, or by a ribbon or a decorative strip of paper. The use of silk cords was the Chinese method but when the cords broke, ribbons or paper were often substituted. The rivet at the base of the sticks which hinges the fan had a washer on either side; these washers could be quite elaborate and some that survive from this period are in the form of a silver or silver-gilt filigree rosette.

The painted decoration on these tiny fans adopted the same subjects as other folding fans of the period: classical motifs, classical or biblical scenes (for example, 'The Finding of Moses', plate 43), a landscape or conversation piece, or a scene from a popular play were all quite common. Usually, the design would completely fill the leaf on the front, while the reverse might bear a few vignettes, though on some fans the reverse traces the scene from the front in a roughly

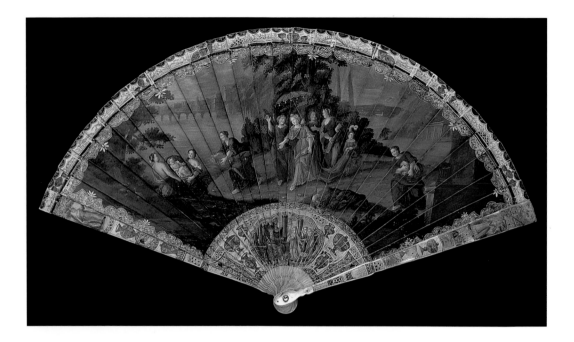

painted outline in red or blue. The painting was in gouache or pure watercolour and invariably of a high standard – indeed comparable to miniature painting in technique and attention to detail. In the early eighteenth century many of these fans were glazed by a high-quality clear varnish which not only provided protection but also enriched the colours, giving them added depth and luminosity. Other forms of decoration included complex and skilled piercing of the fan sticks until they acquired the appearance of lace or filigree with solid areas left for tiny painted vignettes.

The popularity of European brisé fans continued until the 1730s. By this time the fashion for them had declined, as fan makers were taking advantage of the increased availability of paper for fan leaves. Brisé fans remained in decline until the end of the eighteenth century. Their revival at the turn of the century heralded new techniques and a greater variety and availability of materials. Complete fans were made not only of ivory but also horn, mother-of-pearl, tortoiseshell, bone

Plate 43 'The Finding of Moses', ivory brisé fan. Watercolour, varnished. Dutch or French, 1700-20. T.75-1956.

Fig.17 Sandalwood brisé fan. Pierced, applied with prints. English, 1790-1800.
2288-1876.

Fig.18 Reverse of *fig.17*.

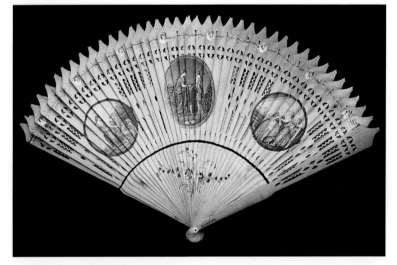

and fragrant sandalwood. Never before had such a variety been accessible to an ever-widening public.

The decoration of these new fans reflects the neo-classical style of the period. One innovation was the introduction of applied prints in the form of vignettes. These were contemporary engravings which had been carefully sliced to fit each stick and yet had to retain the integrity of the original image (fig.17). Not only that, but the fan illustrated is unusual in that it is reversible and capable of showing two different printed designs on each side, depending on whether the fan is opened to the left or to the right (fig.18). There was a naive or childish fondness in the late eighteenth century for this and other types of discreet 'articulation' or puzzle, to be incorporated in a fan to turn a sophisticated dress accessory into an amusing toy.

In the early nineteenth century brisé fans became very small indeed, conforming to the demands of fashion. They were the epitome of contemporary style, which combined elegance with great delicacy. Small ivory or crape fans were carried as evening dress accessories and are constantly referred to in the descriptions of the fashion plates in *Ackermann's Repository*: 'September 1812, "Evening dress, a white crape robe, with demi-train...fan of white and gold crape, or carved ivory"'.

Many brisé fans of this period retained neo-classical motifs

Plate 44 Ivory brisé fan. Gouache with flowers. French, 1810-20.
Circ.246-1953.

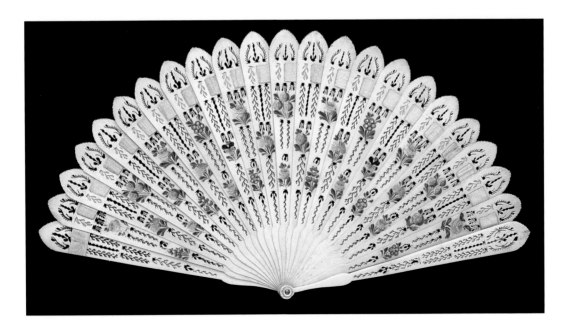

as decoration on the leaf and guard sticks, or kept to a minimalist style; among the most popular designs were small, finely painted floral motifs scattered across the leaf. The sticks were usually carved and pierced, or just pierced, as can be seen in plate 44. A carved ivory fan refered to in *Ackermann's Repository* could have been a small brisé of the type illustrated, or an imported brisé fan from China. Chinese brisé fans were virtually indistinguishable from European fans. Very often there was a central medallion or cartouche which might have a small painted scene or be carved in the form of someone's initials. The initials and painted scene were probably added at a later date, once the fan had been acquired. The fan makers would have originally left that area blank. The East India Company continued to import large numbers of finely carved ivory and tortoiseshell brisé fans from China, made specifically for the western market.

Plate 45 Horn brisé fan. Gouache. English or French, 1820-30s. T.71-1956.

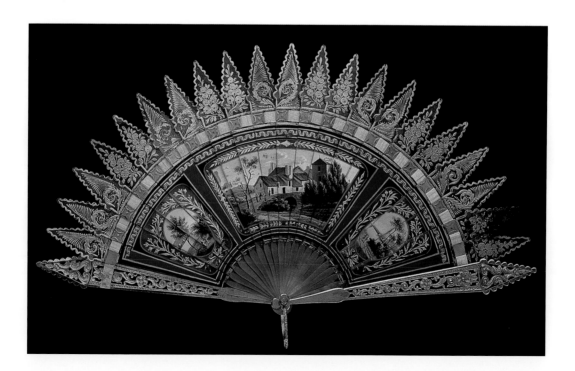

The Gothic Revival, a peculiarly British movement which had its origins in the eighteenth century and took the form of an idealized interest in medieval art and architecture, gathered momentum in the early nineteenth century. Spurred on by the tremendous popularity of the romantic novels of Sir Walter Scott (1771-1832), young women aspired to look like his heroines, such as Amy Robsart. Fashions began to acquire historical overtones. The most striking of these was the so-called 'Van Dyck' style, which took the form of dentated trimmings on collars and cuffs, recalling seventeenth-century styles. On brisé fans it took the form of delicately pointed tops to the sticks which were further embellished with crocketing, inspired by the gables and spires on gothic churches (plate 45).

Other fans of the period remained unembellished as their design relied entirely on pierced decoration, as can be seen in the incredibly delicate ivory fan of 1820-30s (plate 46). This was probably a sample or exhibition piece made to

Plate 46 Pierced ivory brisé fan.
French, 1820-30s.
2282-1876.

85

demonstrate the skill of the craftsman as well as the delicacy of the material he worked with.

Feather fans appear during the nineteenth century but were not much in evidence until the last quarter of the century. Before then imitations of feathers were occasionally used for the shape of the individual sticks, as seen in the fan of *c*.1820-30s in plate 47 and fig.19. This has silk leaves brightly painted in gouache with flowers, exotic birds and insects, and the sticks are of painted and gilded horn.

Plate 47 Silk brisé fan. Gouache with horn sticks.
French or Dutch, 1820-30s.
T.120-1920.

Fig.19 Closed view of *plate 47*, showing feather shape.

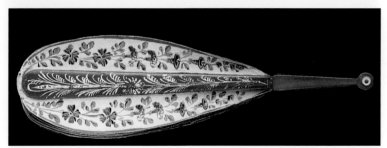

A very fine ivory brisé fan, carved and pierced so effectively that it appears like filigree, is probably of French workmanship *c.*1820-30s (plate 48). This particular example formed part of a collection of fans which originally belonged to Dona Manuela de Rosas de Terrero, the only daughter of General Don Juan Manuel de Rosas (1793-1877) who was dictator of Argentina from 1835-52, and became known in England as the Cromwell of Argentina. On the reverse of the fan is the neat inscription 'Viva La Federacion 24 Mayo', which presumably refers to the federalist government imposed by General de Rosas in 1829-32. This collection of fans was donated to the museum by Dona Manuela's son, Manuel de Terrero, after her death.

Plate 48 Ivory brisé fan. Carved, pierced and painted. French, 1820-30s. T.90-1915.

CHAPTER IX

NINETEENTH-CENTURY AND EXHIBITION FANS

The nineteenth century is probably one of the most exciting periods as far as types and variety of fan production are concerned. The advent of the great international exhibitions helped to create important venues which in turn offered opportunities for improved designs and techniques in fan making. There were also significant changes in women's fashionable dress and this too acted as a stimulus in fan design. The periods of fashionable change can be divided roughly into four periods: 1800-25, 1825-40, 1840-70 and 1870-1900. The fans produced in each of these periods are in keeping with the styles, colours, trimmings and sometimes even the types of textiles in vogue. Their size is also relevant to the fashionable silhouettes of the times.

Between the years 1800-25 fans were, for the most part, small and delicate. This includes the tiny brisé fans of ivory, mother-of-pearl, horn, wood or bone, and the extremely fragile and elegant silk gauze fan leaves (plate 49), appliquéd with delicate gilt paper shapes and supported on daintily decorated sticks of mother-of-pearl, inset with ivory.

Fans became larger and more ornate during the years 1825-40; many are printed and brightly coloured like the group of Spanish ladies on the covers of this book. This particular fan also offers the unusual sight of both front and back views of fashionable dress of the period (c.1830), with a token male in attendance. Other fan subjects are taken from popular literature, songs and plays as well as topical or commemorative themes.

The printed lithographic fan, with watercolour washes and elaborate sticks of mother-of-pearl, ivory, bone, lacquered

wood or papier maché, became the most popular type from about 1840-70 (plate 50). Produced in France and exported in great numbers to other European countries, especially Spain, England, and North and South America, the choice of subject on these fans was usually romantic. Most often the scenes showed a group of people dressed in the style of either the seventeenth or eighteenth century. The richly painted fan in plate 51, on the previous page, is an unusually fine example.

The period 1870-1900 included numerous international exhibitions, where fans of very high quality were displayed and gained high honours. The attraction of historical subjects continued in this period too, but with a difference - romantic, pastoral subjects were chosen from paintings by well-known eighteenth-century artists and copied accurately onto the fan. Plate 52 shows 'The Picnic Party', painted after François de Troy (1645-1752) by F. Lamy.

The extraordinarily large sleeves of the 1890s seem to

have been echoed by the huge fans which appeared at this time. When open, some spanned approximately 30 inches (76cms). These fashions were in complete contrast to the slim figures and tiny fans of the early nineteenth century. Yet by the end of the century, and into the early twentieth century, fashionable dress began to look back to the early 1800s and to re-create the simple and elegant styles of that period. This resulted in the production of poor imitation fans, particularly the small brisé type and those with gauze leaves decorated with silver and silver-gilt spangles. The sticks were usually of bone or wood with machine-tooled decoration, while the spangles were of polished and cut steel.

The mechanization of fan production in England gathered momentum in the nineteenth century, with the result that the quality of design and construction decreased in proportion to the increase of mechanization. Nevertheless, during this period a number of innovative ideas were explored which resulted in some exciting and attractive fans - for not all fans came from the production line.

The use of machine embroidery on fan leaves was an innovation of the second half of the nineteenth century. Machine embroidery was first shown at the Great Exhibition of 1851 in London, by the English firm of Henry Houldsworth of Manchester. The quality of machine embroidery was soon able to compete with hand embroidery, and by the 1870s was used to considerable effect on dresses. A delightful fashionable feature of the period was the notion of producing a fan that matched an outfit (plates 53 and 54). The fan seen here is of English manufacture, with a good-quality machine-embroidered leaf of polychrome flowers on a ground of ivory satin, made to match a machine-embroidered ivory satin ensemble of the 1880s. Sadly the accompanying outfit, which was originally the height of fashion, has been completely unpicked and all that remains are several skirt panels and part of the bodice, giving a frustrating glimpse of how handsome and striking the whole ensemble might once have looked.

As the century progressed the quality of fan leaf designs,

Plates 53 and 54
Machine-embroidered
fan with matching
dress panel.
English, 1880-90.

T.327&A-1965.

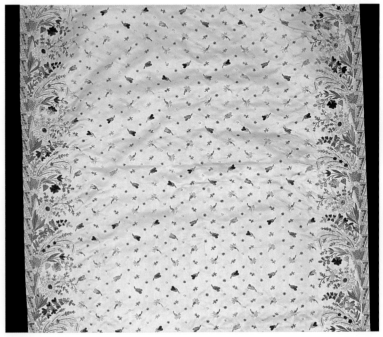

in particular, declined. The techniques and skills still existed
to produce high-quality fans, such as the isolated instances
of exhibition fans or when making the most costly fans, but
even these might attribute their value more to any jewels and
precious metals incorporated in the decoration than to fine
manufacture. The machine-tooling of fan sticks, especially
those made of wood or of bone, was perfected by Alphonse
Baude in 1859. By the end of the century other manufacturers had taken up this technique and improved it, speeding
up fan production. But it was precisely this 'improvement'
which brought about a stereotyped form of design for the
majority of fans, producing sets of sticks which were virtually indistinguishable from one another. Combined with poor
leaf design and painting skills, this only served to reduce the
fan's status as an object of artistic merit. It had simply
become a mass-produced object.

This degeneration caused concern in art circles in
England, and people of influence endeavoured to create
exhibitions and awards to further the improvement of fan
design. One such exhibition, held in 1870, was celebrated
with a catalogue that included twenty photographs of fans
lent to the exhibition by eminent persons. It was the third
show in a series that had started in 1868, organized by the
Department of Science and Art for the Art Instruction of
Women. Prizes were offered for fan designs in a competition
arranged in conjunction with each exhibition. The 1870
exhibition included the prize-winning fan of the year before,
and Queen Victoria also let it be known that she would offer
a fan prize for the forthcoming International Exhibition of
1871. In his introduction to the 1870 exhibition catalogue,
entitled *Fans of all Countries*, Samuel Redgrave (1802-76)
makes the rueful observation that in France fan design and
manufacture is of a particularly high standard, which he contrasts with that of England where 'the trade has not found
such a development, and its future extension would seem to
depend upon the uprising here of men of taste and capital
who, as producers and sellers, shall occupy the place of the
Paris "Eventailliste"'.

Redgrave's acknowledgement of French superiority in fan design may also be attributed to France's own longstanding tradition of major exhibitions of contemporary arts and industrial design. The great nineteenth-century international exhibitions, serving to advance commerce and design, provided the opportunity for fan manufacturers to submit their finest examples for public display. Originating in France in the 1790s, the idea of an exhibition to stimulate interest in various trades and boost the economy was the inspiration of the Marquis d'Aveze, Commissioner of the factories of Sevrès, Savonnerie and of the Gobelin. The principles set out for the first of these exhibitions, held in Paris in 1798, revealed even at this point that there was an awareness of the need to maintain the quality of design – principles which were followed by all subsequent exhibitions. Exhibitors were particularly encouraged to compete for prizes which were offered for quality of design or improved manufacture or technique. The prizes were judged and awarded by a jury. Eleven such exhibitions were held in France between 1798 and 1849, but were only open to French competitors. The first International Exhibition was held in London, in 1851.

Duvelleroy, the firm of fan manufacturers established in 1827 by Jean-Pierre Duvelleroy, was by the second half of the century immensely successful. Establishing an international market, the firm won awards at every major national and international exhibition from 1834 onwards. Virtually dominating the market during the nineteenth and early twentieth centuries, it had shops in the most fashionable areas of London and Paris, with an extensive export market in both North and South America. Its range of designs was wide, catering for the rich and discriminating client as well as for the average and more popular tastes. This was reflected in the prices, which ranged 'from one halfpenny to two hundred pounds', according to the catalogue of the 1862 International Exhibition in London. A number of Duvelleroy fans were displayed in this exhibition and were praised in the catalogue:

Fig.20 Iron brisé fan. German, 1862.
5369-1901.

These fans are frequently decorated by renowned artists, and sometimes contain pictures of rare value, while special attention is paid to the mounting in order to combine grace with strength. They occasionally copy from the finest production of old times, but more usually issue such as have the recommendation of novelty...There are few...sovreigns and princes of Europe who have not been 'customers' to them...we are mainly indebted for [their] perfection...rivalling those that have descended as legacies from our great grandmothers.

The V&A acquired a number of exhibition fans in the second half of the nineteenth century. One most unusual fan, that was manufactured in Germany, was exhibited in the International Exhibition of 1862 in London. This is a brisé fan made of iron (fig. 20) and was one of a pair. They were made under the direction of Edward Schott at Ilsenburg-am-Harz to demonstrate the fineness which could be achieved in iron castings. Both fans were executed in a revival of the early nineteenth-century Gothic style. The other fan of the pair was presented to Princess Frederick William of Prussia (1840-1901), Queen Victoria's eldest daughter.

Most of the exhibition fans in the V&A collection are, however, of French design and manufacture. The superb brisé fan shown in plate 55 came from the 1867 Paris Universal Exhibition. It was made by the Parisian firm of

Plate 55, overleaf Ivory brisé fan, painted by Edouard Moreau (1825-78). Alexandre. French, 1867.
729-1869

729-69

Plate 56 Handscreen, one of a pair. Printed scraps on gauze, white silk fringe. Carved and pierced ivory handle. French or English, 1870s.
T.147-1980.

Alexandre, which consistently produced high-quality fans, competing with Duvelleroy to win the highest honours at the international exhibitions. Fans made by Alexandre and exhibited in the 1867 Exhibition were commissioned by the Empress Eugénie, the Queen of Spain, and by the Empress of All The Russias. The fan shown here is plain on one side with elaborate scenes of a medieval jousting tournament on the other, meticulously painted by Edouard Jean-Baptiste Moreau (1825-78). Edouard Moreau often worked for the fan maker Alexandre and was noted for his finely painted minia-turist compositions which generally represented romanti-cized scenes from the medieval or Renaissance periods. The painted vignettes are separated by standing figures repre-senting, from the left, Magnaminité, Charité, Justice and finally Concorde. The sticks are of moulded ivory composi-tion, and the fan is signed 'Alexandre' in red paint. Each vignette is signed by Moreau. Nearly all the V&A's exhibition fans came from the workshops of Alexandre.

Whether any of these exhibitions had the desired effect of producing good design thereafter is debatable. The exhibi-tions certainly displayed some very fine fans, but they were, in a sense, one-off examples made specifically for show. They do not appear to have made much impact on the aver-age manufactured product: aimed at the general public, the majority were fans characterized by insipid designs which were easy on the eye and the pocket.

Another type of fan which should be mentioned here is the rigid fan, or 'handscreen' as it became known in the eighteenth and nineteenth centuries. Generally used indoors beside the fire, they may also have been used outside as pro-tection from the sun. As their name might suggest, hand-screens served to protect or screen ladies' complexions from fireside heat. The fans were usually produced in pairs, with one placed on either side of the mantlepiece in readiness. Decoration on these fans was varied: the pretty example from the 1870s in plate 56 has an elaborate handle of carved and pierced ivory, and its gauze ground, trimmed with a white silk fringe, is randomly covered with highly coloured printed 'scraps' depicting flowers, birds and people. These paper decorations could be bought by the sheet, and were intended for decorating furniture such as the larger folding screens. This early form of Do-it-yourself was an activity which had become a popular pastime for ladies. It is quite possible that this handscreen - one of a pair - was bought ready-made, but undecorated.

CHAPTER X

FANS OF THE BELLE ÉPOQUE IN ENGLAND AND FRANCE

A plate in the fashion magazine *The Queen* for 10 August 1901 shows elegantly dressed ladies chatting after dinner in the Palm Court of the Carlton Hotel in London. All are carrying fans, but one in the foreground appears too exhausted to talk and sits alone, fanning herself with a large painted silk fan. The focus of attention is an impossibly slim, glamorous woman who has just walked in, carelessly resting her half-open fan on her shoulder. Images like this show the importance of the fan to the fashionable woman at the turn of the century. Fig.21, from *The Woman's World* (February 1890), also shows what an essential accessory to evening dress fans had become. One woman carries a feather fan of a style much in fashion from the 1880s to the 1920s, while another has opened out her painted silk fan.

It seemed that fans were back in fashion again after a period of decline. *The Artist* of 1 August 1885 commented that 'fashion has turned her inventive genius to fans'. How different is this remark from the depressing observation in *Whittock's Complete Book of Trades* of 1842 that 'fans are now to be obtained at the Haberdasher's Shops, but are now so little used as to be nearly or quite obsolete'. The large numbers of high-quality examples surviving from the mid-nineteenth century are evidence that this was not in fact the case, and fans were exhibited at all the major international exhibitions. But there was certainly a renewal of popular interest in fans as the century drew to a close. An attractive but optional accessory in the middle of the century, they had become by 1890 an essential part of any fashionable wardrobe, and were made in an unprecedented range of

Fig.21 Plate from *The Woman's World*. Detail. February 1890.
III RC E 17 - NAL.

shapes, sizes, materials and styles. The huge sleeves and flamboyant, flowing lines of turn-of-the-century fashionable dress seemed ideally matched by large, dramatic fans.

All the fans in this chapter date from the period known as the *belle époque*. This is a term that conjures up the cultural excitement and eclecticism prevailing in France, and especially Paris, in the decades preceding the First World War. The café-concerts, the singers and dancers of Montmartre immortalized by Henri Toulouse-Lautrec (1864-1901), and the glittering cafés and shops on the *grands boulevards* are all inextricably associated with the *belle époque*. In relation to fashion and fans, the term describes the exceptional elegance and extravagance of fashionable dress from 1885 to 1914.

While the British experience of this period is different, and there is no equivalent phrase in English, there are some useful points of comparison. Above all, a sense of recklessness and a feeling that the old world was ending infected the final years of the nineteenth century. A fan by the great French fan painter Maurice Leloir expresses this poignantly. On one side is 'Le Nouveau Siècle', a beautiful woman riding on a mythological bird, while on the other a mermaid-like creature with huge wings and flowing tail, 'Le Siècle Finissant' (the century ending) is raising her wings for the last time as she slips into the ocean waters. Also significant is the pre-eminence of Paris as a fashion centre at this time. Most wealthy Englishwomen bought their dresses and fans in Paris, or from Paris firms with London shops, like the fan maker Duvelleroy.

The V&A has a stronger collection of English than French fans for this period. One exception is shown in plate 57, the quintessential turn-of-the-century Parisian fan, sumptuously decorated and coloured. However, while Parisian fan painters developed styles linked to high fashion and the work of commercial and theatre artists, English fans of the

Plate 57 Painted in watercolour on silk, with coloured mother-of-pearl sticks. French, *c.*1900. T.60-1970.

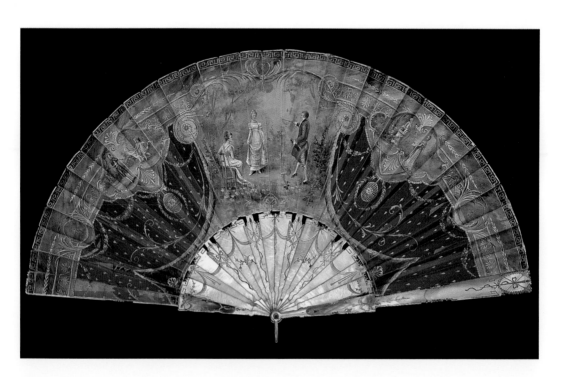

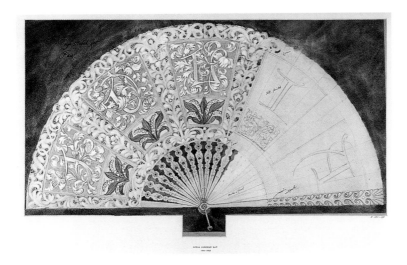

Fig.22 Fan design by
Lewis F. Day.
English, 1888.
E. 1021-1911.

period often reveal more complex influences.

In London, even avant-garde artists such as Aubrey
Beardsley were heavily influenced by the Arts and Crafts
movement, and students at art school were encouraged to
design in the Arts and Crafts tradition. Many produced fans,
some of which were illustrated in the leading art journal, the
Studio. Moreover, some of the leading figures of the Arts and
Crafts movement designed and painted fans. The 1888 fan
design illustrated in fig.22 is signed and annotated by Lewis
F. Day (1845-1910), a founder member of the Arts and Crafts
Exhibition Society, and one of the most versatile and commercially successful designers of his generation. From 1870,
he produced freelance designs for textiles, wallpaper, ceramics and furniture, and from 1888 he exhibited designs for
fans, lace and embroideries. This example is described as a
'rough sketch for a fan', and indicates that it should have a
lace leaf, with inset painted silk panels, each spelling a letter
of the name 'Julia', and mother-of-pearl sticks with real pearls
on the guardsticks. The stylized lilies of the valley painted on
the silk panels are particularly typical of Day's work.

Phoebe Traquair (1852-1936) also worked in an astonish-
ing range of techniques, including mural painting, embroi-

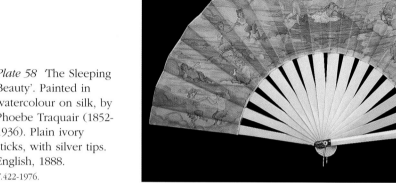

Plate 58 'The Sleeping Beauty'. Painted in watercolour on silk, by Phoebe Traquair (1852-1936). Plain ivory sticks, with silver tips. English, 1888.
T.422-1976.

dery, manuscript illumination and enamelling. Based in Edinburgh all her life, by 1900 she was Scotland's foremost artist of the Arts and Crafts movement. In 1906 the poet W.B. Yeats met her, and praised her 'extraordinary abundance of imagination. She has but one story, the drama of the soul'. Indeed most of her work has a mysterious religious quality. Yeats also mentions her 'childlike' charm, and the fan in plate 58 with its fairytale title, 'The Sleeping Beauty', and its plump cherub figures certainly has a childlike air. It is painted on silk, signed and dated 1888, the same year as Day's fan design. Embroidered fans were also produced at this time by Ann Macbeth and other designers associated with the Glasgow School of Art.

Well-known painters sometimes tackled fans, too. George Sheringham (1884-1937), later to become a successful painter and theatre designer, held his first solo exhibition in London in 1910, exhibiting silk panels and fans. The V&A has several of his fan leaves, including the one in fig.23. Sheringham's work was influenced by Japanese art, and his fan leaves include several oriental scenes. This example, entitled 'The Cobweb Fan', is partly a stylish pastiche on an eighteenth-century fan, with figures in eighteenth-century dress in three vignettes. The striking monochrome 'cobweb' background, however, is much more contemporary, and reminiscent of Aubrey Beardsley's work.

The Sheringham fan leaf is the closest in this chapter to an Art Nouveau fan. Many French fans from about 1900 borrow elements of Art Nouveau, especially the asymmetrical and sinuous treatment of flowers. Alphonse Mucha (1860-1939), a leading light of the Art Nouveau movement, did design one handscreen fan in 1899, entitled 'Le Vent qui passe emporte la Jeunesse' (the passing breezes take youth with them). Gendrot (first name unknown, active *c*.1900), another designer with a typically Art Nouveau style, designed many fans for Duvelleroy, including 'La Muse de Champagne' (Champagne's muse) in 1900.

Despite the evident involvement of leading artists in painting and designing fans, it was felt by many that the quality of fan painting was declining and that too few talented artists were involved. In October 1902, Madeleine Lemaire, a leading French fan painter, wrote an article for *Fémina* lamenting the fall in standards. She was perhaps unduly pessimistic. While the leading establishment artists honoured in the Paris Salon may have disliked fan painting, both Degas and Gauguin, among other artists more celebrated now than in their own time, painted superb fan leaves.

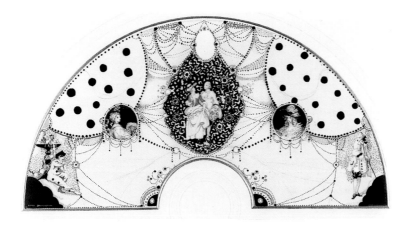

Fig.23 'Cobweb Fan'.
Fan leaf by George
Sheringham.
English, 1910-20.
E.142-1914.

107

Plate 59 Hunting scene,
signed on the reverse
by the artist, Marie,
Duchess of Orléans.
French, c.1885.
T.38-1957.

Fig.24 Detail from
reverse of plate 59.

Amateur painters also made a contribution to the art of fan making. Fan painting was regarded as a respectable accomplishment for young women of good family. Innumerable advertisements in magazines of the 1880s and 1890s announce new paints and materials for fan painters. *The Artist* for 1 Jan 1885 advises that 'Fans, painted on Satin and Silk in Oil Colours are very effective and will not crack if painted with the medium "the Adolphi Process"...Sold in bottles, with instructions, 1s 6d'. Madeleine Lemaire's 1902 article gives detailed instructions about the types of paper and vellum available, ready prepared, for the use of fan painters.

There were some well-known royal fan painters. Queen Victoria's daughter Princess Louise painted fans, and Marie, Duchess of Orléans, a member of the French royal family, was a talented artist who specialized in painting hunting scenes on fans. Plate 59 shows a good example of her work. The huntsman in red, the dogs and the rider have been skilfully fitted into the fan leaf shape, and the woodland setting

Plate 60 Gouache on silk, with applied mother-of-pearl. French for the English market, *c.*1900.
T.263-1972.

is cleverly suggested using a light, impressionistic technique. On the reverse (fig.24), the name 'Marie' is painted in a flowing script, so the artist may have intended the dashing rider to be a self-portrait. From 1885 until the late 1890s fans with hunting scenes were at the height of fashion, many coming to Paris from Austria.

The most admired professional fan painters of the *belle époque* worked for the main Parisian fan makers such as Duvelleroy and Alexandre. By 1887 Duvelleroy had two shops in Paris and one in London, employing celebrated fan painters such as Madeleine Lemaire and Louise Abbéma. Using the eighteenth-century tradition of coded gestures to publicize his own firm, Duvelleroy published a leaflet called 'The Language of the Fan', a list of messages and their meanings which included 'Drawing across the cheek' (I love you) and 'Twirling in the right hand' (I love another). The messages, while certainly flirtatious, have always caught the imagination of those interested in fans.

Plate 60 almost certainly shows a Duvelleroy fan. The subject and clothes are a deliberate echo of those on eighteenth-century fans, but the green-stained mother-of-pearl sticks

Fig.25 Detail from reverse of *plate 60*.

and the rounded shape of the leaf are typical of fans from around 1900. The principal charm of this fan is the way the image on the front is seen from behind on the reverse. A girl talks to her lover over a fence while her father looks on, horrified. On the reverse (fig. 25), the lover reaches up to talk to his sweetheart. This device was very popular with English clients, and fans employing it are known to have been made by Duvelleroy in Paris exclusively for sale in London. The V&A has another similar example.

The fan in plate 61 is signed by Ronot-Tutin, a French fan painter who specialized in floral fan leaves. The top part of the leaf is painted on silk gauze, with a French bobbin lace panel underneath, and mother-of-pearl sticks. The pansies are extremely realistically painted, though their size and colour suggest a certain vulgarity. This fan may have been made for an English customer, or someone who wanted a fan to match a very particular dress.

Not all fan painters of the *belle époque* were French, as we have seen. Francis Houghton and his daughter Rosie were

Plate 63 Gouache on textured paper. English or French, *c*.1910.
T.265-1972.

prolific and talented English painters. Very little is known about them, but many of their fans survive. Francis Houghton's fans tend to depict girls or couples in eighteenth-century style dress, in woodland settings. Plate 62, signed by Houghton, shows a wilder and more dramatic scene than is usually associated with him. The man leaning over the side of the boat is presumably saving the girl, who is stranded out at sea in a rowing boat. Her beauty is reminiscent of the work of the Pre-Raphaelites, unlike the pretty rosy-cheeked faces often painted on fans of this date. The fan is large, at over two feet wide certainly the biggest in this book, and the sea has been painted to give the impression of stretching into infinity, even spilling onto the sticks, a common device in English fans of the 1890s. There is also a fan by Rosie Houghton in the V&A collection. In First Empire style, with painted vignettes on a black ground, it demonstrates that the 1900 passion for Napoleonic fans was not exclusively French.

Plate 63 shows what might be described as a novelty fan, dating from the early years of the twentieth century. With its

leaves cut in the shape of a pigeon in flight, it may appear slightly grotesque, but was one of many fans of the time which used the shapes of cats' heads, birds or butterflies to form the fan leaf.

The leading fashion designers of the early 1910s, Paul Poiret (1879-1944) and Madame Paquin (1869-1936) among them, were searching for new influences for their designs. Diaghilev's Ballets Russes, touring Europe from 1908, had introduced a variety of oriental fashions through their extravagantly costumed performances. In June 1911, Poiret gave a fancy dress party called the '1002nd Nights'. Poiret supplied appropriate 'Persian' dress for his guests, and himself appeared as the Sultan. Harem pantaloons were worn by many of the women, and although they outraged many people, a version of them remained in fashion. Paquin worked with the fashion illustrator Georges Barbier to develop a series of fans on an Eastern theme. Plate 64 shows one of this group, dated 1911. The stylized grapes framing the scene and repeated on the painted sticks appear on several Barbier fans. The girl wears harem pants, and a headdress identical to the one worn by Mme Poiret at the '1002nd Nights' party. This fan is in part an advertising fan for Paquin – it would have had smart 'Paquin' packaging – but was also a highly desirable fashion accessory. Dating as it does from the end of the *belle époque*, in style and content it looks forward to the 1920s and to Art Deco.

Plate 64 Georges Barbier for Paquin. Printed and hand coloured on paper, with painted silk reverse. Painted bone sticks. French, 1911. T.333-1978.

Postscript

Since 1914, the story of European fans has been essentially one of decline. In the early twentieth century many high-quality fans were still produced, some by leading artists of the age. In the 1920s extravagant feather fans were considered the perfect accessory for a slimline evening dress, but since then fans have enjoyed only fitful popularity, and expensive, finely crafted examples are extremely rare.

A painted silk fan by Duncan Grant (plate 65) dates from around 1913, and properly belongs to the previous chapter, but in style it is perhaps the most obviously modern fan in the V&A collection, and provides the most startling contrast with the finely painted late seventeenth-century narrative fans in Chapter 1. Although it does not contain the characteristic mark of the Omega Workshops – the Greek letter 'omega' – this fan is very similar to surviving examples painted for them by Duncan Grant. The Omega Workshops, founded in June 1913, were the brainchild of Roger Fry, and several other members of the so-called 'Bloomsbury Group' were involved. The Omega style is best represented by the contents of the home of Vanessa Bell and Duncan Grant, Charleston in East Sussex. It is on publicity material for an Omega Workshops exhibition in November 1913 that Omega fans are first advertised, and this fan must date from around that time.

Another twentieth-century artist to make a contribution to the art of fan painting was Oskar Kokoschka (1886-1980) who painted a series of fans in 1913-14 as a tribute to Alma Mahler, the wife of the composer Gustav Mahler.

Fans of the 1920s were either tiny – versions of the minute brisé fans of the 1820s – or enormous. Embroidered and

sequinned examples were common, as were coloured feather fans. The French writer Colette (1873-1954), in her story 'Châ', describes one character as wearing 'a silver dress, with sulphur-yellow roses at the waist, a large fan of sulphur-coloured feathers, and her hair, skilfully bleached to a very pale yellow, looked like some sort of finery bought at the same time as the roses and the fan'.

Certain types of fans retained their popularity. A significant number of lace fan leaves were exhibited by English ladies at the 1925 Paris International Exhibition, and the fan shape has remained popular with lacemakers. Embroiderers have also used fans to demonstrate their design ability and technical skills. Mary Kessell, a leading embroiderer working for the Needlework Development Scheme in the 1940s, designed a fan now in the V&A and fans are still occasionally created by contemporary embroiderers.

Among the first advertising fans were the 1880s examples

Plate 66 Fan advertising the Brasserie Universelle, Paris. French, 1920s. T.257-1978.

Fig.26 Reverse of *plate 66*, with photographs of the restaurant.

produced in the Far East for the Paris department store Les Grands Magasins du Louvre. From the 1890s until the Second World War shops, restaurants and theatres all produced cheap printed paper fans to advertise their services. The popular decorative style of the day – whether Art Nouveau or Art Deco – inspired the designs.

The V&A has few advertising fans in its collections, but plate 66 shows an excellent Art Deco example of one. On the reverse of this striking Pierrot fan (fig.26) are photographs of the Brasserie Universelle restaurant in Paris. The series of fans from Parisian department stores were espe-

cially remarkable. Printemps commissioned a different fan
each year from the 1900s to the 1930s, often from a leading
illustrator, and each with the slogan 'Toute Femme Elégante
est cliente du Printemps' (Every Elegant Woman is a customer of Printemps). In 1990, on the occasion of the shop's
125th anniversary, Printemps paid tribute to this long-dead
tradition, producing a fan whose leaf showed a photograph
of the remarkable stained glass in their shop on Boulevard
Haussmann. A comprehensive collection of Printemps fans is
held by the Musée de la Mode et du Costume, Paris.

Fashion designers might seem to be the most obvious
people to promote their work with fans, and indeed many
did. Poiret and Paquin were mentioned in the last chapter,
and plate 67 shows a 1950s fan by the most influential
designer of the immediate post-war period, Christian Dior.
The design on the leaf and the materials used could not be

more simple, but the asymmetrical shape of the leaf is a clever and sophisticated touch. Interestingly, in their 1985 show Dior included simple woven straw handscreen fans, examples of which are also in the collection of the Musée de la Mode et du Costume. They would surely have been appreciated by Dior himself.

It is above all as souvenirs of special occasions that new fans are now designed and made. In January 1973 a gala evening was held at the Royal Opera House, Covent Garden, to celebrate Britain's entry into the EEC. The evening was titled 'Fanfare for Europe' and a printed card fan, with details of the programme, was given to guests. Fans are distributed at catwalk shows by contemporary designers – recent examples include Hermès and Louis Féraud – although these are often Chinese or Japanese fans. The designer Karl Lagerfeld invariably carries a fan at his own catwalk shows.

Fans will never lose their appeal. Indeed in hot countries like Spain, fan shops selling hundreds of designs can still be found. However, the close relationship between the history of European fans and the development of dress and the decorative arts which lasted from the seventeenth to the early twentieth century seems to have ended. Fans are either produced cheaply and commercially, or as one-off curiosities to commemorate an event or to display the maker's skill in lace making or embroidery.

While it is hard to believe that such a functional object will ever die out entirely, every lover of fans must hope that leading artists and designers will one day turn their attention once again to creating these beautiful accessories.

GLOSSARY

Articulated fan Fan with one or more moving parts, usually on the guardstick and activated by a movable pin.

Brisé fan Fan with no leaf, composed entirely of sticks held by a rivet at the base and joined by a ribbon along the top edge.

Découpé fan Fan with a vellum or paper leaf, cut out or stamped out to make a decorative, lacy pattern.

Folding fan Fan with a continuous pleated leaf, mounted on sticks.

Gorge Area of the sticks above the rivet and below the leaf of a folding fan.

Gouache Opaque watercolour medium, which uses pigment mixed with white gum and other thickeners.

Grisaille Painting in shades of grey, using only black and white pigment. Often associated with mourning fans.

Guardsticks The outer sticks, usually heavier and more elaborate than the inner sticks. They protect the fan when closed.

Mica A semi-opaque mineral, consisting of thin scales, which can be used in small pieces to decorate fans.

Piqué Type of inlaid decoration, using silver or steel dots, set in ivory, mother-of-pearl or tortoiseshell sticks.

Rivet The pin holding the sticks together at their base.

Sticks Skeleton or framework of a fan, consisting of guardsticks and narrower inner sticks.

Vellum Fine leather, specially prepared for writing on or painting.

Major Museums with Fan Collections

Great Britain: Fan Museum, Greenwich,
                   London
                   British Museum, Department of
                   Prints and Drawings, London
                   Fitzwilliam Museum,
                   Cambridge
France: Musée de la Mode et du
                   Costume, Palais Gallièra, Paris
                   Musée Carnavalet, Paris
The Netherlands: Rijksmuseum, Amsterdam
Germany: Bayerisches National Museum,
                   Munich
USA: Metropolitan Museum of Art,
                   New York
                   Museum of Fine Arts, Boston

Acknowledgements

The authors would like to thank Paul Harrison for his valuable additional research and advice on the text. We are grateful to V&A Publications for guiding the book through publication, in particular to Mary Butler, Miranda Harrison and Ariane Bankes. We also thank Valerie Chandler for her index.

We gratefully acknowledge the help of Valerie Mendes and Amy de la Haye from the Textiles and Dress Department of the V&A.

Finally, we are very grateful to Sara Hodges for her exquisite photography, which was specially commissioned for this book and executed with great skill and precision.

Selected Bibliography

Armstrong, Nancy. *A Collector's History of Fans*, London, Studio Vista, 1974.

Arnold, Janet. *Queen Elizabeth's Wardrobe Unlock'd*, Leeds, Maney, 1988.

Bennet, A.G. and Berson, R. *Unfolding Beauty: The Art of the Fan*, exhibition catalogue, Museum of Fine Arts, Boston, 1988.

Cust, Lionel. *Catalogue of the collection of fans and fan leaves presented to the Trustees of the British Museum by Lady Charlotte Schreiber, 1893.*

Delpierre, M. and Falluel, F.(eds.) *L'Eventail: Miroir de la Belle Epoque*, exhibition catalogue, Musée de la Mode et du Costume, Palais Gallièra, 1985.

Gostelow, Mary. *The Fan*, Dublin, Gill and Macmillan, 1976.

Impey, Oliver. *Chinoserie*, Oxford, Oxford University Press, 1977.

Mayor, Susan. *Collecting Fans*, London, Christies/Studio Vista, 1980.

Rhead, George Woolliscroft. *History of the Fan*, London, Kegan Paul; Trench, Trübner & Co., 1910.

Schreiber, *Lady Charlotte. Fans and Fan Leaves, vol.1: English; vol.2: Foreign*, London, British Museum, 1888-90.

INDEX